Instrumental Methods
for Rapid
Microbiological
Analysis

Instrumental Methods for Rapid Microbiological Analysis

Edited by Wilfred H. Nelson

Wilfred H. Nelson, Ph.D.
Professor of Chemistry
University of Rhode Island
Pastore Chemical Laboratory
Kingston, Rhode Island 02881

Library of Congress Cataloging-in-Publication Data
Main entry under title:

Instrumental methods for rapid microbiological analysis.

   Includes bibliographies and index.
    1. Diagnostic microbiology.   2. Instrumental
analysis.   I. Nelson, Wilfred H., 1936-
QR67.I56   1985   616.07'58   85-15726
ISBN 0-89573-137-1

© 1985 VCH Publishers, Inc.

Printed in the United States of America.

ISBN 0-89573-137-1   VCH Publishers
ISBN 3-527-26315-2   VCH Verlagsgesellschaft

PREFACE

The last quarter century has seen a revolution in the application of sensitive, rapid, and increasingly informative methods of chemical analysis. This has happened to a large extent because advances in electronics, computers, and laser technology have allowed the practical application of methods that previously had been understood in theory but were cumbersome to use.

The list of these methods is a long one, ranging from atomic and molecular spectroscopy, on one hand, to separations and surface science. In addition to infrared and nuclear magnetic resonance spectroscopy, which began the revolution, we must add the various fluorescence spectroscopies, Raman and resonance Raman spectroscopy, and associated techniques, such as coherent anti-Stokes Raman spectroscopy (CARS). Separation methods, such as gas chromatography (GC) and high pressure liquid chromatography (HPLC), coupled with mass spectrometry (MS) or alone, have produced their own revolution. These and many other methods have increased vastly our capacity to do basic research. Yet, they have had an even greater impact on the industrial laboratory, making previously difficult analyses relatively cheap and routine and providing many opportunities for automation.

During this same 25 years, progress in the microbiology laboratory, although impressive, has been slower. At present, because of enormous advances in molecular biology, the foundation has been made for much more rapid progress. This progress is inevitable because, now that the molecular biology of microorganisms is better understood, it is time that many aspects of the earlier revolution in the chemistry and physics laboratory be applied directly to the study of living cells. I believe one fruitful area to be the development of rapid methods for the identification of microorganisms.

At present, many effective methods for the identification of microorganisms exist. These vary widely in sensitivity and specificity. Most commonly, they involve cultural studies of organisms, which require much time and often lack specificity unless extremely tedious procedures are followed. The lack of speed characteristic of most routine analysis is a serious problem. Frequently, important decisions concerning the presence of pathogens have to be made before the results of microbiological tests are available.

In response to this problem, a number of attempts have been made to apply instrumental methods to rapidly characterize microorganisms. Methods described in this book will have a major impact on both the clinical laboratories and the industries exploiting advances in biotechnology. It is my hope that this volume will be helpful both to chemists and to biologists who labor in this extraordinarily productive area.

CONTRIBUTORS

Alvin Fox, Ph.D., Department of Microbiology and Immunology, School of Medicine, University of South Carolina, Columbia, SC 29208

Mack Fulwyler, Ph.D., Department of Laboratory Medicine, University of California, San Francisco, San Francisco, CA 94143

W. Keith Hadley, Ph.D., Department of Laboratory Medicine, University of California, San Francisco, San Francisco, CA 94143

K. A. Hartman, Ph.D., Department of Biochemistry and Biophysics, University of Rhode Island, Kingston, RI 02881

Richard W. Hutchinson, Ph.D., Chemical Systems Laboratories, Edgewood Arsenal, MD 21010

Stephen L. Morgan, Ph.D., Department of Chemistry, University of South Carolina, Columbia, SC 29208

Harold A. Neufeld, Ph.D., United States Medical Research Institute of Infectious Diseases, Fort Detrick, Frederick, MD 21701

Judith G. Pace, Ph.D., United States Medical Research Institute of Infectious Diseases, Fort Detrick, Frederick, MD 21701

Thomas M. Rossi, Ph.D., Department of Chemistry, Emory University, Atlanta, GA 30322

Mahadeva P. Sinha, Ph.D., Jet Propulsion Laboratory, 4800 Oak Grove Drive, Pasadena, CA 91109

G. J. Thomas, Jr., Ph.D., Department of Chemistry, Southeastern Massachusetts University, North Dartmouth, MA 02747

Frederic Waldman, Ph.D., Department of Laboratory Medicine, University of California, San Francisco, San Francisco, CA 94143

Isiah M. Warner, Ph.D., Department of Chemistry, Emory University, Atlanta, GA 30322

D. M. Yajko, Ph.D., Department of Laboratory Medicine, University of California, San Francisco, San Francisco, CA 94143

CONTENTS

1. BACTERIAL IDENTIFICATION USING

FLUORESCENSE SPECTROSCOPY

Thomas M. Rossi and Isiah M. Warner

I. INTRODUCTION

The current interest in the application of instrumental techniques to bacterial identification results largely from the limitations of more traditional identification procedures. Classical bacterial identification procedures usually include a morphological evaluation of the microorganism as well as tests for the organism's ability to grow in various media under an assortment of conditions, susceptibility to various phages and antibiotics, and ability to metabolize various compounds. Generally, no single test provides a definitive identification of an unknown bacterium. Hence, a complex series of tests is often required before an identification can be confirmed. The results of such a series are often difficult to interpret and not available on the time scale desired in the clinical laboratory.

These considerations have prompted a number of workers to explore modern instrumental alternatives to classical identification procedures. Modern analytical instruments are characterized by rapid acquisition and high reproducibility of data and frequently feature computer-aided data recording and interpretation. A number of unique instrumental approaches have been applied to the identification of bacteria. This chapter is a review of fluorescence spectroscopic methods for bacterial identification and provides a brief review of other instrumental methods that have been explored, such as gas chromatography and infrared spectroscopy.

A. Gas Chromatography

A number of studies have explored the use of pyrolosis-gas chromatography (PGC) for detection and identification of microorganisms [1]. Whole cells, cell fractions, and metabolic products of bacteria have been investigated using PGC. Pyrograms of microorganisms contain predominantly amino acids, sugars, carbohydrates, purine and pyrimidine bases, etc. In short, the pyrograms contain the basic molecular building blocks of life. The first applications of PGC to the study of microorganisms were aimed at developing a life detection system to be used in the search for extraterrestrial life [2,3]. The earliest work in identification of bacteria by PGC was reported by Reiner in 1965 [4].

The pyrograms of microorganisms and cell fractions are generally very complex. Frequently, microorganisms must be differentiated by two or three characteristic peaks in a pyrogram containing two hundred or more components [5-9]. In the past, this has meant tedious visual comparisons of many complex pyrograms. Recently, however, the development of computerized pattern recognition techniques has helped to speed up the data analysis process [10-12].

Gas chromatography alone, ie, without pyrolysis of the bacterium, also has been used to identify bacteria. Because whole cells cannot be examined directly by gas chromatography, most of the work in this area has involved the detection of specific metabolic byproducts [13-15]. Alternatively, volatile compounds have been extracted from bacteria and analyzed by gas chromatography [16].

B. Infrared Spectroscopy

The use of infrared (IR) spectra as a means of identifying bacteria was first reported in the 1950s [17-20]. The basic procedure is simple. Bacteria are smeared onto an IR cell and IR absorbance spectra are acquired using conventional instrumentation. The resultant spectra represent the chemical composition of the bacterium under investigation and, hence, are generally very complex and broad band because of the large number of components in the spectra. It is often true that identifications must be made based on minor differences between complex spectra of many bacteria. Also, questions have been raised as to the reproducibility of the data. Because of the difficulties and inherent limitations of this technique reports of its application to bacteria became less frequent in the 1960s and virtually ended in the mid 1970s. The availability of Fourier transform techniques may revive interest in the IR spectra of bacteria.

C. Fluorescence Spectroscopy

The main limitation of both the gas chromatographic (GC) and IR approaches to bacterial identification is that both techniques involve an evaluation of the chemical composition of bacteria. Bacteria are, on the whole, remarkably similar at the molecular level, although they are chemically very complex. Hence, the inherent chemical nature of bacteria has forced users of GC and IR identification methods to search for minor differences in complex data sets. In contrast to these approaches, fluorescence-based identification techniques do not generate data from all the individual molecular components of the microorganism. Also, there is a very broad diversity of fluorescence spectroscopic identification procedures, including detection of microorganisms, metabolic products, and fluorescent tags. Because of the diversity of these methods, it is convenient for the organization of the remainder of this chapter to classify fluorescence-based identification techniques as being either primary or secondary methods. Each of these classifications is subdivided into direct and indirect methods.

Primary fluorescence methods are those in which the natural fluorescent components of the bacterium are examined. An example of a primary fluorescence method is the identification of <u>Bacteroides</u> species by the fluorescence of cells held under an ultraviolet lamp [21]. Some species of <u>Bacteroides</u> were found not to fluoresce, whereas others emitted fluorescence of characteristic colors. There is at least one major limitation to the utility of primary techniques. That is, only those bacteria which contain or produce some fluorescent pigment may be examined by primary techniques.

Secondary fluorescence methods, in contrast, involve the introduction of a foreign fluorophore to the bacteria before identification is made. These methods do not depend on any natural fluorescence and therefore can be applied more generally. Hence, secondary methods are by far the most useful. Immunofluorescence assays (IFA) can be placed in this category of techniques [22, 23].

Both of these categories can be further subdivided into two more general classifications, direct and indirect methods. Direct analyses are those in which the bacterial cell is present during the analysis. Indirect methods are those in which the cell is not present during the fluorescence analysis. Examples of direct and indirect techniques are flow-cytometric measurement of DNA content and detection of fluorescent pigments in cellular growth media, respectively.

This chapter discusses the variety of fluorescence techniques that have been used for bacterial identification. The range of levels of sophistication of these techniques is broad. Special attention is given to the most sophisticated of these studies, ie, those techniques which utilize new and innovative fluorescence instrumentation for the identification of bacteria. Because the most advanced work in this field is also the most recent, the future potential of these methods, as well as the present level of development, is evaluated. In order to appreciate fully the power and potential of these new identification schemes, it is necessary first to review the basic theory of molecular fluorescence. For this reason, the main body of this chapter begins with a section on the theory and instrumentation of fluorescence spectroscopy. Primary importance is placed on instrumental developments that have a high potential for use in bacterial identification.

II. THEORY AND INSTRUMENTATION OF MOLECULAR

FLUORESCENCE SPECTROSCOPY

A. The Fluorescence Process

Most molecules at room temperature in fluid solution exist in the ground (lowest energy) electronic and vibrational energy states. For most organic molecules, this is a singlet state, designated S_0. To initiate the fluorescence process a molecule first must be excited from the ground state to a higher energy state. This discussion explores the use of photons for excitation, although many forms of excitation are possible. Once the molecule has been excited it must release energy to return to (relax to) the ground state. In fluorescence, this relaxation process proceeds via emission of a photon from the excited-state molecule. This sequence of events can be described using the equations:

$$A + h\nu_\alpha \quad \rightarrow \quad A^* \tag{1-1}$$

$$A^* \quad \rightarrow \quad h\nu_f + A \tag{1-2}$$

where A is the ground-state molecule, A^* is the excited molecule, h is Planck's constant, and ν_α and ν_f are the frequencies of the absorbed and emitted photons, respectively.

The process of absorption and fluorescence, as well as alternate modes of relaxation, are illustrated in the Jablonski diagram in Figure 1-1. Note from this diagram that the molecule is promoted to a higher quantum energy level within 10^{-15}S after initiation of the absorption process. The energy of the absorbed photon must exactly match the energy difference between the ground state and the excited state. Regardless of the particular excitation transition (as long as photoionization is not induced) the molecule generally relaxes to the lowest vibrational level of the S_1 (first excited) state via internal conversion before fluorescence occurs. Fluorescence, then, usually occurs only from the S_1 state; hence Eq. (1-2) can be rewritten:

$$S_1 \quad \rightarrow \quad h\nu_f + S_0 \tag{1-3}$$

because Eq. (1-3) is independent of the excitation transition, the fluorescence spectrum of a molecule is usually independent of the excitation energy. Hence, two generally independent types of spectra can be obtained for a fluorophore: the excitation spectrum and the emission spectrum. This general rule is particularly important in later discussions of fluorescence methods in mixture analysis.

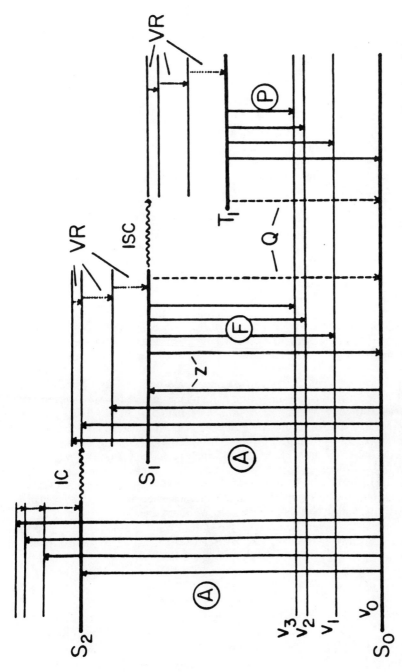

Figure 1-1. Jablonski diagram illustrating: absorption transitions, A; fluorescence transitions, F; phosphorescence transitions, P; internal conversion, IC; intersystem crossing, ISC; vibrational relaxation, VR; and quenching, Q, where S_0, S_1, S_2 and T_1 are electronic energy levels and v_0, v_1, v_2, and v_3 are vibrational energy levels.

It is also evident from Figure 1-1 that fluorescence is not the only means by which an excited-state molecule can relax back to the ground state. Another alternative involves intersystem crossing to the T_1 state followed by a nonradiative or radiative transition down to the S_0 state. The radiative process of phosphorescence may be described by:

$$S_1 \rightarrow T_1 \rightarrow S_0 + h\nu_p \qquad (1\text{-}4)$$

where ν_p is the frequency of the photon emitted as phosphorescence. Radiative transitions between energy states of different multiplicity are quantum mechanically forbidden and, therefore, phosphorescence lifetimes are usually in excess of 10^{-4} s: whereas fluorescence generally occurs in about 10^{-8} s. Also, phosphorescence occurs at longer wavelengths than fluorescence.

At this point, it is useful to define the quantum efficiency of fluorescence, ϕ_f:

$$\phi_f = \frac{\text{number of quanta of energy emitted}}{\text{number of quanta of energy absorbed}} \qquad (1\text{-}5)$$

Because fluorescence is not the only possible mode of relaxation, it follows that the ratio of photons emitted (through fluorescence) to photons absorbed is not likely to be unity. The quantum efficiency of a fluorophore is dependent on the competition of fluorescence with the other relaxation pathways.

Competition between fluorescence and the other relaxation processes also can be used to define another quantity, the mean radiative lifetime, τ_r, of the excited state. All of the relaxation processes diagrammatically illustrated in Figure 1-1 follow first-order kinetics. Hence, each can be assigned a first-order rate constant. However, one additional competing process needs to be discussed, ie, collisional quenching. Collisional quenching is a second-order kinetic process. However, its rate constant usually is defined as the product of the quenching constant, K_q, and the concentration of the quencher, $[Q]$, because the required collisional quencher concentration, $[Q]$, is usually sufficiently large that quenching may be considered a pseudo-first-order process. Therefore, the quantum efficiency of a fluorophore can be considered to depend on the rates of the relaxation processes according to the following relationship

$$\phi_f = \frac{K_f}{K_f + K_{IC} + K_{ISC} + K_q[Q]} \qquad (1\text{-}6)$$

where K_f, K_{IC}, and K_{ISC} are the rate constants for fluorescence, internal conversion, and intersystem crossing, respectively. Similarly, the mean radiative lifetime of the excited state is defined as

$$\tau_r = \frac{1}{K_f + K_{IC} + K_{ISC} + K_q [Q]} \tag{1-7}$$

Time-resolved fluorescence techniques can be used to resolve individual spectra from a mixture based on differences in τ_r between fluorophores.

B. Quantitative Fluorescence Measurement

One of the primary advantages of fluorescence over absorption spectrophotometry is its relatively low detection limit. This advantage has helped make fluorescence a desirable method for many types of analysis, including bacterial identification. In the present section of this chapter the quantitative aspects of both absorption and fluorescence are discussed. The interrelation of the two techniques is explored, as is the reason for the greater versatility of fluorescence measurements.

The quantitative relationship between the absorbance of radiation and the concentration of a chromophore is given by the Beer-Lambert law:

$$A = \log P_0/P = \varepsilon bc \tag{1-8}$$

where A is the absorbance, P_0 is the power of the incident radiation, P is the power of the radiation after it has passed through the sample, ε is the molar absortivity, b is the path length, and c is the concentration of the chromophore. In order for the Beer-Lambert law to be applicable the absorbance must be small. Hence, the measurement of A involves determining a small difference between two large signals. This is not a desirable condition for sensitive analytical measurements. By conventional measurement techniques the lower limit of measurable absorbance is usually on the order of $A = 10^{-3}$. Because the ratio P_0/P is not affected by a change in P_0, it is not possible to increase the sensitivity of the absorbance measurement by increasing P_0. In contrast to absorbance, the power of fluorescence is dependent on the magnitude of P_0. The power of fluorescence is dependent on the amount of absorbed radiation, P_α, where P_α is defined as:

$$P_\alpha = P_0 - P = P_0 (1 - 10^{-\varepsilon bc}) \tag{1-9}$$

By considering Eq. (1-8) and (1-9), it is evident that whereas an increase in P_0 has no effect on A, it does cause an increase in P_α. Because the power of fluorescence of a sample is dependent on P_α, it is apparent that the sensitivity of a fluorescence measurement can be increased by increasing P_0. At low absorbance, the equation describing the power of fluorescence has the form:

$$P_f = \phi_f P_0 \, 2.3 \, \varepsilon bc \tag{1-10}$$

For solutions where $A \leq .01$ Eq. (1-10) is quite accurate. An additional factor contributing to the sensitivity of fluorescence is the nature of the measurement process itself. In the measurement of fluorescence, P_f is measured relative to a blank contribution, usually near zero. This is inherently a much more accurate process than the measurement of A.

This discussion has provided a brief introduction to the theory of fluorescence spectroscopy. For a more thorough discussion of this topic the interested reader is referred to any of several good sources [24-27].

C. Conventional Fluorescence Instrumentation

A block diagram of a generalized fluorescence spectrometer is shown in Figure 1-2. Instead of depicting a specific fluorometer, this diagram represents the basic components common to most fluorometers. The instrumentation discussed in the remainder of the chapter can be considered in terms of Figure 1-2 by placing specific components in the "black boxes." Some common components of fluorometers are discussed below.

The light source provides photons for excitation of the sample. The xenon arc lamp is a common light source. It has a continuous (although not flat) output in the ultraviolet (UV) and visible wavelengths. Another light source used in some of the instrumentation to be discussed is the laser. The unique properties of laser radiation that make it useful for a fluorescence light source have been reviewed [28,29].

In most fluorometers the excitation beam is passed through a wavelength selector before it reaches the sample. The wavelength selector typically rejects all but a narrow range of wavelengths and allows a nearly monochromatic beam to reach the sample. Various types of filters and diffraction gratings have been employed as monochromators. Instruments that use diffraction gratings can be used to scan a range of wavelengths. In such instruments an emission spectrum may be acquired by scanning a range of emission wavelengths while monitoring the fluorescence intensity at a constant excitation wavelength. Similarly, a range of excitation wavelengths may be scanned while the monitored emission wavelength is held constant to produce an excitation spectrum. Filter-based instruments, on the other hand, usually cannot scan wavelength ranges.

Conventional fluorometers usually monitor only one wavelength of emission from the sample at any one time. Hence, an emission monochromator generally is placed between the sample and the detector. The same types of wavelength-selecting devices used for excitation are also used for emission wavelength selection.

The photomultiplier tube (PMT) is the typical detector used in conventional fluorometers. The PMT is sensitive to UV and visible light. It is probably the most sensitive type of detector currently available for UV radiation. Other detectors that are sometimes used, such as the photodiode array and the vidicon, will not be discussed here.

The geometry of the instrument depicted in Figure 1-2 is typical of fluorometers. The emission beam is monitored 90° relative to the excitation beam. Other configurations sometimes are used when sample concentration is very high or when a unique optical arrangement can help reduce interference from scattered light.

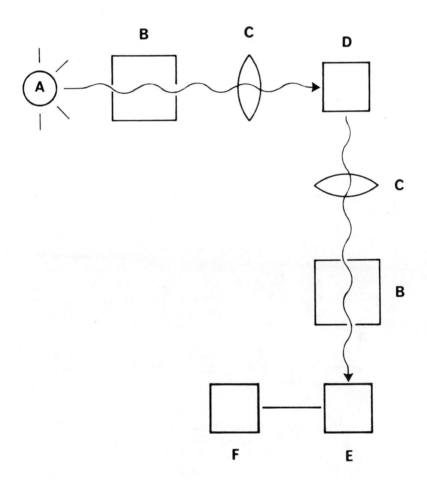

Figure 1-2. Schematic representation of a generalized fluorescence spectro-photometer where A is the excitation source, B are wavelength selection devices, C is the optics, D is the sample holder, E is the detector, and F is the readout device.

Fluorometers of the general type discussed above are typically very sensitive instruments. Analyses of strongly fluorescing compounds typically can be made at concentrations as low as 10^{-10} M. However, as a general rule fluorescence spectra are acquired at one excitation and emission wavelength pair at a time. Hence, to record an emission spectrum spanning 200 nm at 1nm resolution, at least 200 separate data points must be recorded individually.

If the dependence of the emission spectrum of a given sample on the excitation wavelength is to be evaluated, the process of acquiring the emission spectrum must be repeated at as many excitation wavelengths as are desired. This makes most fluorometers poor tools for examining mixtures of fluorophores if the number and identity of the components are not known prior to the analysis.

D. Examination of Mixtures of Fluorophores

To derive analytically useful information from a mixture of fluorophores, it is often necessary to resolve the spectrum of each component in the mixture. There are essentially three solutions to this problem. One possibility is to use a sample pretreatment method, such as chromatography, to separate the components physically prior to spectroscopic analysis. However, such methods can be undesirable because of the sample loss, sample changes, and increased analysis time associated with additional sample-handling steps. Two other approaches are time-resolved fluorescence and total luminescence spectroscopy (TLS) [30].

Time-resolved spectroscopy depends on each component in the mixture having a unique radiative lifetime. The theory and practical aspects of time-resolved fluorimetry have been reviewed thoroughly [30,31], but a brief discussion is given here. The rapid process of fluorescence decay ($\sim 10^{-8}$ s) can be measured by the time-correlated single-photon (TCSP) technique. In TCSP the time between excitation of the sample with a nanosecond pulse of radiation and the detection of the first emitted photon is measured using a sophisticated electronic system. After a large number of replicate measurements, a histogram of time after excitation versus number of occurrences of the first detected photon is constructed. The shape of the histogram yields information about the radiative lifetimes and number of components in a mixture.

Knorr and Harris recently have demonstrated the use of fluorescence decay information acquired at various emission wavelengths to resolve binary mixtures of fluorophores [32]. This method required no a priori information about the number and identities of the components. Although time-resolved techniques have proved useful for mixture analysis, they have not yet been applied to bacterial identification. However, as the technique matures applications to bacterial identification are likely to be explored.

In contrast to TCSP, total luminescence spectroscopy (TLS) is a mixture analysis technique that has found applications in bacterial identification. This technique is discussed in detail because of its relevance to the subject of this chapter. The instrumentation and theory of TLS is discussed in this section and the data reduction techniques used in TLS are discussed in the next section.

The instrumentation of TLS ranges in complexity from computer-controlled conventional fluorometers [33-36] to the video fluorometer (VF) [37-39]. These instruments are all used to collect data in a two-dimensional matrix known as an emission-excitation matrix (EEM) [40]. However, these instruments may require over 30 min. to acquire an EEM over a similar spectral range and with similar resolution as the VF. Because of the time advantage of the VF most of the recent work in TLS has been done using this instrument.

The rapid data acquisition capabilities of the VF arise from a unique optical configuration and detection system. Figure 1-3 is a schematic representation of the video fluorometer. Notice that whereas conventional fluorometers use monochromators as wavelength selectors, the VF uses polychromators. The polychromators disperse the incident radiation into its component wavelengths. However, unlike the monochromator, a polychromator allows a broad range of wavelengths of light, all spatially separated from each other, to exit the wavelength selector. In the VF the excitation and emission polychromators are orthogonally positioned as with conventional fluorometers. Excitation wavelengths are dispersed along the long axis of the sample cuvet, as illustrated in Figure 1-4. The emission from each excitation band is then dispersed in a direction 90° rotated in the same plane. This results in a two-dimensional image, also depicted in Figure 1-4. Recording this image requires the use of a two-dimensional detector. The VF utilizes a silicon-intensified target (SIT) vidicon to transduce this image to an electronic signal [41-43]. The signal is then shipped to the controlling minicomputer via a high-speed parallel interface [44] and stored as a two-dimensional array on a mass storage device. Typically, EEMs are stored as 50 x 50 or 64 x 64 arrays. Each element of the array contains the fluorescence intensity of the sample at a unique excitation and emission wavelength pair.

Although it is evident from the preceeding discussion that an EEM contains much more information than a conventional one-dimensional spectrum, it may not be apparent why this is a particularly useful approach to mixture analysis. A consideration of some theoretical aspects of fluorescence will lead to an understanding of mixture analysis using TLS. It was stated earlier in this chapter that regardless of the specific excitation transition, fluorescence emission generally occurs according to Eq. (1-3) An argument was developed which indicated that for a single fluorophore the shape of the emission spectrum was generally independent of the excitation energy and vice versa. During the following discussion the importance of this rule of thumb to TLS mixture analysis will become evident.

T. M. Rossi and I. M. Warner

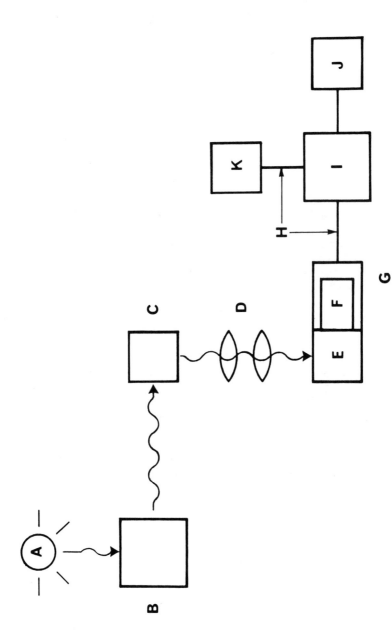

Figure 1-3. Schematic representation of the video fluorometer, where: A is a xenon arc lamp, B is the excitation polychromator, C is the sample cuvet, D represents the emission optics, E is the emission polychromator, F is a silicon-intensified target vidicon, G is a cooled housing for the detector, H represents data transfer lines, I is a multichannel detector controller, J is a real-time monitor, and K is the controlling minicomputer.

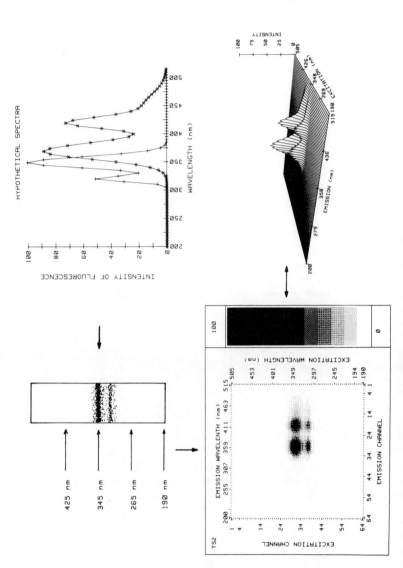

Figure 1-4. Illustration of the generation of an EEM from a hypothetical fluorophore. (a) Hypothetical emission and excitation spectra, (b) cuvet image, (c) monochromatic representation of the image at the point of detection, (d) isometric projection of resultant EEM.

It logically follows from the above discussion that if the shape of the emission spectrum of a sample does exhibit a dependence on the excitation energy, the sample generally must contain more than one fluorophore. The same argument also applies to the dependence of the shape of the excitation spectrum on the monitored emission wavelength. An example of these principles is provided in Figure 1-5. Notice that as the excitation wavelength is varied, the emission profiles for pure anthracene and for perylene are attenuated but not changed in shape. The emission spectrum of the mixture, however, does change in shape as the excitation wavelength is changed. It is also important to note that by a judicious choice of excitation wavelengths, either component of the mixture may be excited selectively. If the identities of the components in the mixture are known prior to the analysis it becomes easy to choose the proper wavelengths for selective excitation. However, if the identities of the components are not known prior to analysis it may be necessary to examine the emission spectra acquired at a number of different excitation wavelengths before selective analysis is possible. An EEM can be considered to be such a collection of emission spectra. The EEMs of the perylene, anthracene, and a mixture solution are shown as isometric projections in Figure 1-6. From these spectra, it is immediately evident that the mixture contains at least two components, whereas the other two spectra are probably single-component samples. It is also evident that the interdependence of the emission and excitation spectra of a sample can be evaluated at a variety of wavelengths by examining a single EEM. It is this feature of the data, combined with the rapid data acquisition capacity of the VF, which makes TLS a convenient method for examining mixtures of fluorophores [45-51]. Many advanced mathematical methods have been developed for extracting information from EEMs [401]. A complete description of these is beyond the scope of this chapter. However, a brief review of these techniques is warranted so that the reader may appreciate more fully the power of TLS and its potential use in bacterial identification.

E. Data Reduction Methods

Data reduction methods for TLS began to appear in the literature in the late 1970s and active research is still being conducted in this field. Warner et al reported the application of eigenanalysis to EEMs in 1977 [52]. This technique was useful for qualitative evaluation of the data. Later in that same year Warner et al reported quantitative analysis of EEMs by the methods of least squares and nonnegative least sum of errors [53]. Eigenanalysis was adapted later by Ho et al to yield quantitative information [54,55]. In the early 1980s a ratio method for the resolution of multicomponent EEMs was reported by Fogarty and Warner [45,46]. Recently, Rossi and Warner have developed two-dimensional Fourier transform-based algorithms for filtering [56], quantitative analysis [57], and automatic pattern recognition of EEMs.

The EEM has been referred to previously as a data matrix. For a spectrum of a single fluorescing species, this matrix can be represented by the general equation

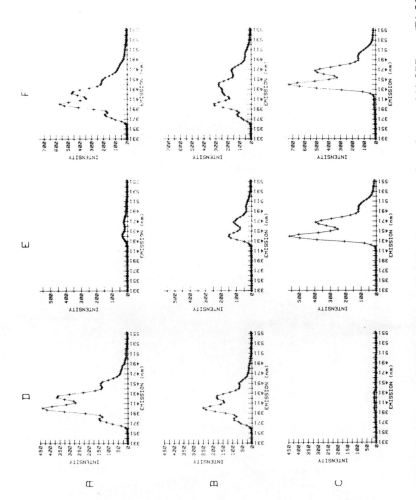

Figure 1-5. Emission spectra of pure and mixture samples. Excitation wavelengths are: (A) 355nm, (B) 360 nm, (C) 378 nm for emission spectra of (D) anthracene, (E) perylene, (F) mixture of anthracene and perylene.

T. M. Róssi and I. M. Warner

$$M = \alpha \ x \ y \qquad\qquad (1\text{-}11)$$

where M is the EEM, α is a concentration-dependent term, x is a column
vector representing the excitation spectrum and y is a row vector representing
the emission spectrum. The vectors x and y represent the excitation spectrum
and emission spectrum, respectively. In dilute solutions of fluorophores,
synergistic effects, such as energy transfer and self-absorption, usually can be
ignored. Hence, the EEM of a mixture of fluorophores can be treated as a

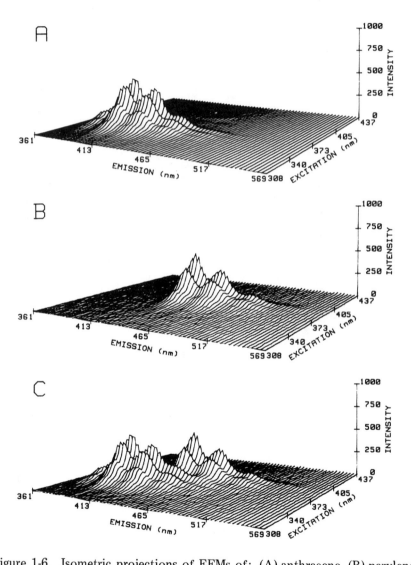

Figure 1-6. Isometric projections of EEMs of: (A) anthracene, (B) perylene,
and (C) an anthracene and perylene mixture.

linear combination of the EEMs of each pure component. This can be expressed in algebraic form:

$$M = \sum_{k=1}^{r} \alpha_k \, x^k \, y^k = \sum_{k=1}^{r} M_k \tag{1-12}$$

where k denotes the kth component and M is the EEM of the mixture containing r components. These simple principles provide the foundation for discussing the individual data reduction schemes.

Eigenvector analysis has been used to approximate the rank, r, of EEMs and to isolate each M_k for a two-component system. In an ideal, noise-free EEM the rank of the matrix is equal to the minimum number of fluorescent components in the sample. The success of the eigenvector analysis approach is dependent on the noise content of the data as well as on the degree and type of overlap between components. For an EEM containing random noise, it has been found that the calculated value of r equals the dimension of the matrix. No definitive method has been developed to determine the true number of components in an EEM. Present criteria for determining the number of eigenvectors representing spectral information sometimes disregard minor components in a sample. Also, the derived M_k are sometimes distorted by high degrees of spectral overlap in the mixture EEM. Because of these limitations, eigenvector analysis generally is applied to mixtures containing only two or three components if the components are spectrally similar. As an example, Fogarty and Warner have applied eigenvector analysis to photochemical studies [58].

The extension of eigenanalysis to provide quantitative information about a sample is based on a rank annihilation strategy. The method calls for subtracting a standard, γM_k, from M and adjusting the factor, γ, until the rank of M is reduced by exactly one. In mathematical terms, when the equation:

$$M^r - \gamma M_k = M^{r-1} \tag{1-13}$$

is satisfied, the concentration-dependent term, γ, has been estimated.

An alternative method of quantitative analysis using two-dimensional least squares also has been proposed. In this method an error matrix, E, is constructed such that the elements of this matrix are defined by:

$$E_{ij} = M_{ij} - \Sigma \gamma_k \, (M_{ij})_k \tag{1-14}$$

where i and j represent row and column positions. A weighted sum of the squares of the error matrix elements is minimized by estimating γ. This method requires that the number and identities of the mixture components be known prior to the analysis. If the information is not available an extension of this method based on simplex analysis can be used. The interested reader is referred to the original study for further details [53].

Another method for quantitative analysis and resolution of component spectra is ratio "deconvolution." The term "deconvolution" used in naming this technique is a mathematical misnomer. A more appropriate name would be ratio matrix resolution (RMR). A prerequisite for RMR is that for a mixture containing r components, a set of r EEMs be available. In each of these EEMs, the observed concentration factor, γ_k, for each component should change relative to each of the other components in the mixture. Hence, for a binary mixture of A and B the two mixture matrices defined below must be available.

$$M_1 = \gamma_{A1}A + \gamma_{B1}B \tag{1-15}$$

$$M_2 = \gamma_{A2}A + \gamma_{B2}B \tag{1-16}$$

where $\gamma_{A1}/\gamma_{A2} \neq \gamma_{B1}/\gamma_{B2}$. The EEMs M_1 and M_2 then can be ratioed element by element, resulting in a ratio matrix. In general $r-1$ ratio matrices are required to resolve r components. A system of linear equations can be applied to the ratio matrix to determine the relative concentration factors (eg, γ_{A1}/γ_{A2} and γ_{B1}/γ_{B2}) and to resolve the EEMs of each component. The RMR procedure becomes difficult to implement when the noise content of the EEM increases. Also, RMR requires that each component have an area of fluorescence that is free from overlapping fluorescence from any of the other components. This can be a rather severe constraint.

It is evident from the foregoing discussion of data reduction methods that data with low noise are handled more easily than noisy data. A method for reducing the random noise content of EEMs while exerting the minimum effect on spectral information has been proposed recently. This method falls into the general category of Fourier transform-based data reduction schemes. The theory of Fourier transforms is not within the intended coverage of this chapter and is sufficiently complex to preclude a nonmathematical discussion of the topic. The reader interested in Fourier transform theory is referred to either of two pertinent texts [59,60]. For this discussion it is sufficient to say that the Fourier transform is used to convert the EEM to a new mathematical form in which it can be manipulated and inversely transformed to yield new, useful information.

Two manipulation processes have been used with EEMs: convolution and correlation. The convolution operation involves multiplication of the transformed EEM by a filter function followed by inverse transformation to yield a smoothed or otherwise filtered EEM [56]. The application of this mathematical data reduction technique to EEMs is analogous to work previously done on one-dimensional spectra [61,62]. An example of data smoothed by this method is provided in Figure 1-7. Several spectra of 9,10-dimethylanthracene solutions of various noise content are shown before and after smoothing. Fourier transform smoothing is a useful tool for interpreting data acquired near the detection limit of the instrument.

A process that is mathematically similar to convolution is correlation [63,64]. Correlation-based quantitative analysis is particularly useful for the evaluation of noisy EEMs. The method is applicable to both single-component samples and mixture EEMs. Perhaps of greater importance to the subject of this chapter is the application of correlation analysis as a means of pattern recognition for EEMs.

Correlation functions can be considered to be a means of evaluating the likeness of two EEMs. When one EEM is correlated with another, the resultant function is either an autocorrelation or a cross correlation. Examples of typical auto and cross correlation functions are given in Figure 1-8. A general feature of the auto-correlation function is that the maximum is located at the center of the matrix. Cross correlation usually does not result in a centrally positioned maximum. These two functions are distinguished easily by a computer. Correlation of an EEM with an EEM of the same fluorophore results in an auto-correlation function. Correlation of dissimilar EEMs results in the cross-correlation function. The usual strategy for pattern recognition is to develop a library of standard spectra with which an unknown can be correlated. The resultant correlation functions can then be evaluated to determine which is closest to the auto-correlation function. For this method to be effective the unknown must have an appropriate match in the library. Hence, correlation-based pattern recognition may be particularly useful for differentiating an unknown from a small number of possibilities.

The capabilities of Fourier transform-based techniques combined with the other data reduction schemes discussed above constitute a powerful set of mathematical EEM data analysis methods. Each of these methods has a unique set of capabilities and limitations. The choice of which method to use for a given application is based on the type of information desired from the EEM as well as the restrictions imposed by noise content, spectral overlap, and other features of the data. Many of the data reduction methods discussed here may be applicable to bacterial identification based on TLS. Throughout the remainder of the chapter both demonstrated and potential areas of application of video fluorometry to bacterial identification are discussed. It is hoped that the material in this section of the chapter will help the reader appreciate the potential utility of TLS for bacterial idenficiation. It will be seen that many fluorescence spectroscopic identification schemes can be considered as simple mixture analysis problems.

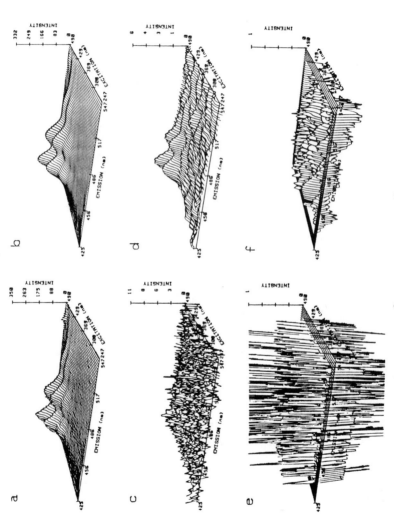

Figure 1-7. Isometric projections of: (a) 2.07 x 10^{-6} M, (c) 8.28 x 10^{-9} M solutions of perylene in cyclohexane, and (e) pure cyclohexane. These same data are shown in (b), (d), and (f), respectively, after application of an optimized smoothing filter.

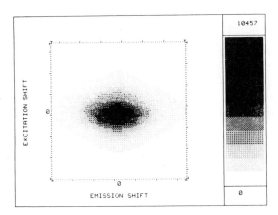

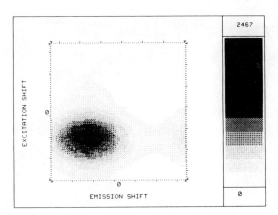

Figure 1-8. Monochromatic images of: (a) an auto-correlation function; and
(b) a cross-correlation function.

III. PRIMARY FLUORESCENCE METHODS

Primary methods of fluorescence analysis are those which involve the detection of some naturally fluorescent component of the bacterium. Generally a mixture of fluorescent metabolic byproducts is detected. Many different procedures have been used to detect the primary fluorophores, ranging in complexity from visual examination of specimens under UV light to complex isolation, chromatography, and spectroscopy schemes. Regardless of the specific analysis procedure chosen, all bacteria examined by primary methods must meet one criterion; the microorganism must produce or contain some suitable fluorophore. Although this seems quite obvious, it must be considered, as it is mainly this criterion that limits the applicability of primary methods.

Some bacteria that have been widely reported to contain or produce fluorescent pigments are listed in Table 1-1. It is likely that many other bacteria exhibit native fluorescence that has not been observed because of the

Table 1-1. Bacteria Containing or Producing Fluorescent Pigments

Bacterium	Reported pigment(s)	References
Bacteroides melaninogenicus spp. asacchrolyticus intermedius melaninogenicus	Not identified	21, 65, 66
Propionibacterium acnes	Uroporphyrin, coproporphyrin, protoporphyrin	55, 67
Pseudomonas aeruginosa P. fluorescens P. mildenbergii P. putida P. tolaasii P. Ovalis	Pteridine, peptide complexes	68 - 76
Rhodopseudomonas sphaeroides, Rps. viridis	Bacteriochlorophyl, bacteriopheophytin, metalloporphyrins	77, 78
Veillonella alcalescens, V. Parvula	Not identified	79

crude detection methods commonly employed and the apparent lack of an organized search for fluorescent bacteria. In many schemes used in the clinical environment, fluorescence is detected visually while the sample is held under a UV lamp. This scheme has the advantages of simplicity, low cost, and rapidity. However, the utility of this approach is also very limited because subtle differences in the fluorescence characteristics of the tested microorganisms are difficult to evaluate. An example of the visual evaluation method is the identification of Bacteroides menaninogenicus spp. by Slots and Reynolds [21]. In that study it was found that specific strains have unique emission characteristics. However, in order to differentiate between strains it would have been necessary to distinguish between such closely matching emission colors as red orange and pink orange. Any identification based on this type of subjective evaluation would be questionable. However, the fact that different colors of fluorescence are reported for such closely related bacteria suggests that the use of more sophisticated instrumental methods may provide valuable "fingerprinting" information. Also, instrumental techniques are likely to be more sensitive than visual evaluation. An increase in sensitivity may enable workers to see smaller quantities of bacteria than presently are needed for analysis, as well as to find bacteria that fluoresce too weakly for visual detection. The complexity of bacteria as well as the reported variety of emission colors suggest that any instrumental technique used must have the capability of performing multicomponent analysis. Hence, it can be seen immediately that TLS may be very useful in improving the quality of the information obtained in primary analyses.

A. Primary Fluorophores Detected by TLS

Recently, Shelly et al demonstrated the utility of the video fluorometer for primary fluorescence fingerprinting of bacteria [75,76]. A list of the Pseudomonas aeruginosa and Pseudomonas fluorescens strains used in this study is given in Table 1-2. Cultures of these bacteria were grown in minimal salts growth media, the components of which are listed in Table 1-3. After a 24-h growth period the cultures were centrifugated and the supernatants filtered to remove any remaining cells. The supernatants were then extracted with ether to isolate fluorescent pigments produced by the bacteria. These isolates were then examined using video fluorometry.

Figure 1-9 is an illustration of the cuvet images of each bacterial extract. Note that excitation bands within individual cuvets were observed to emit light of various colors. This is evidence that the profiles of the emission spectra of these samples depend on the excitation wavelength. Hence, many of these samples may be assumed to contain complex mixtures of fluorophores. Also, the cuvet image of each sample is unique, indicating that the EEMs of these isolates may serve as a useful means of differentiating these bacteria. An examination of the EEMs in Figure 1-10 confirms that each bacterium produces a unique mixture of pigments. However, the EEMs of P. fluorescens FWS and P. aureginosa TDH 0492 are more similar than is predictable from the cuvet images shown in Figure 1-9. The discrepency between

Table 1-2. Pseudomonas Examined by Shelly et al [75,76]

Bacterium	Strain	Sources
P. aeruginosa	ATCC 9721	American Type Culture Collection
	MMI-2	Clinical isolate, TAMU Health Center
	TDH-0492	Texas State Department of Health
	TDH 0461	Texas State Department of Health
	TDH-06089	Texas State Department of Health
	FWS	Dr. Billy Foster, TAMU
P. fluorescens	FWS	Dr. Billy Foster, TAMU

TAMU, Texas A&M University.

Table 1-3. Composition of "Minimal Salts"
Growth Medium Used by Shelly et al [75,76]

Component	Medium (#g/L)
Glucose	2
K_2HPO_4	7
$K H_2PO_4$	3
$MgSO_4 \cdot 7 H_2O$	0.1
$(NH_4)_2SO_4$	1
$NaC_2H_3O_2$	0.5

visual and instrumental evaluation of the cuvet images underscores the importance of using instrumental techniques to record primary fluorescence patterns.
Aside from providing a more accurate and reproducible means of qualitative
analysis, video fluorometry yields quantitative information that is not
obtainable by visual evaluation methods. Quantitative comparisons of the two
EEMs indicated that pigment production was approximately 125-fold greater
in P. aureginosa FWS than in P. fluorescens TDH 0492. In general, both the
quantitative and the qualitative characteristics of the bacteria tested in this

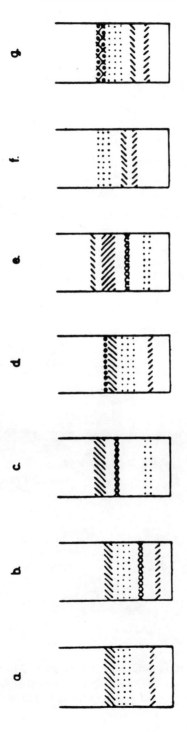

Figure 1-9. Cuvet images of media extracts: (a) P. auruginosa ATCC 9721, (b) P. auruginosa TDH 0492, (d) P. auruginosa FWS, (e) P. auruginosa TDH 0461, (•) P. auruginosa TDH 06084, (g) P. fluorescens FWS. Color codes: (●) purple; (X) dark blue; (///) light blue; (•) greenish yellow, and (\\\\) yellow. (Reprinted with permission from Clinical Chemistry.)

study were found to be reproducible as long as growth conditions were held constant.

Growth conditions have long been known to affect pigment production in pseudomonads. Hence, primary methods, such as the work described above,

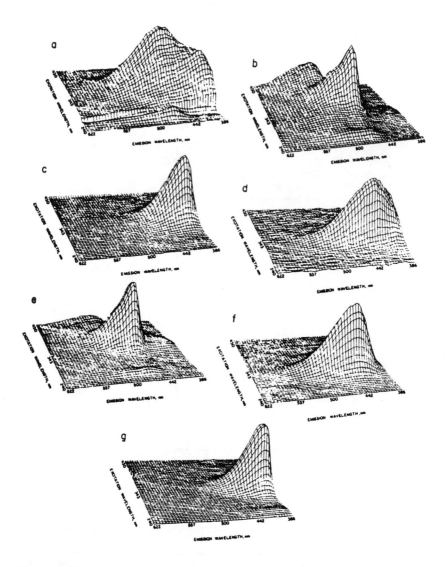

Figure 1-10.   Isometric projections of fingerprint EEMs of P. auruginosa: (a) ATCC 9721, (b) MMI-2, (c) TDH 0492, (d) FWS, (e) TDH 0461, (f) TDH 06089, and (g) P. fluorescens FWS. (Reprinted with permission from Clinical Chemistry.)

are likely to benefit from growth optimization studies [80]. In further work by Shelly et al, the effect of growth conditions on the EEM fingerprinting method was investigated [76]. Such parameters as agitation of cultures, argenine supplementation of the growth medium, aeration, and growth time were varied. As a result of the optimization study the minimum growth time required for the bacteria to produce detectable quantities of pigments was shortened from 24 to 4 h. In addition to optimizing growth conditions, this followup study included the use of chromatography and video fluorometry to aid in the identification of the pigments that contributed to the EEM fingerprint.

As a first step in the pigment identification study, culture filtrates were examined using cation-exchange liquid chromatography. The column effluent was monitored using a conventional fluorometer set at excitation and emission wavelengths of 405 and 450 nm, respectively. Four of the cultures contained pigments in sufficient quantities to be examined by this method. Typical chromatograms of these four culture filtrates are shown in Figure 1-11. The complexity of these chromatograms indicates that a large number of pigments may be contributing to each of the EEMs of Figure 1-10. Further information about each eluent in the chromatogram was needed before the pigments could be identified. Additional spectral information about each eluting pigment was provided by interfacing the HPLC to the VF. Figure 1-12 is a group of isometric projections of EEMs obtained "on the fly" during chromatographic separations of the pigments. In spite of the apparent complexity of each chromatogram, only four spectroscopically distinct pigments were found. The fact that these pigments were observed repeatedly throughout the chromatograms suggests that a few fluorophores may be bound to a variety of macromolecules, possibly polypeptides. If this is true, it is the chromatographic properties of the polypeptides, not the pigments, that give rise to the complex nature of the chromatograms in Figure 1-11.

Shelly et al reached the following conclusions from these studies. First, the bacteria in Table 1-2 can be identified and differentiated based on TLS evaluation of their fluorescent pigments. Second, the fingerprinting properties of the EEMs shown in Figure 1-10 are probably the result of a small number of fluorophores present in various concentrations in each bacteria. Finally, the spectral characteristics of the isolated pigments are consistent with previous speculation that the pigments may be pteridine-type compounds [73,74].

B. Future Prospects in Primary Fluorescence Techniques

Although the studies outlined in Section III.A prove that TLS can expand the accuracy of primary methods greatly, the problem still exists that the number of bacteria in Table 1-1 is small. Clearly, for primary methods to be more widely applicable, the number of bacteria in Table 1-1 must be expanded. Fluorescent components of many bacteria not listed in Table 1-1 may prove adequate for effective application of primary methods using sophisticated instrumentation such as the VF. For example, recent work by Wood indicates that fluorescent c-type cytochromes isolated from bacterial membranes may be useful as a basis for primary identification methods [81].

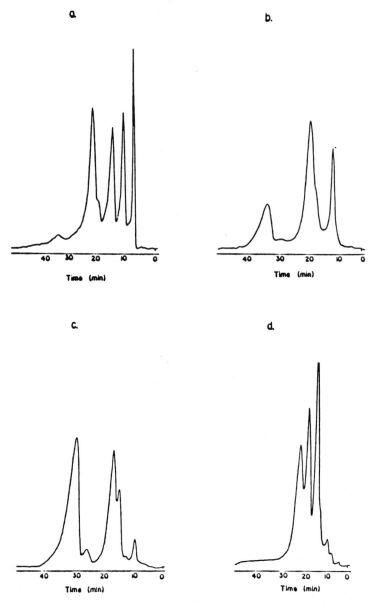

Figure 1-11. Chromatographic profiles of P. aurginosa: (a) ATC 9721, (b) MMI-2, (c) FWS, and (d) TDH 0492. (Reprinted with permission from Clinical Chemistry.)

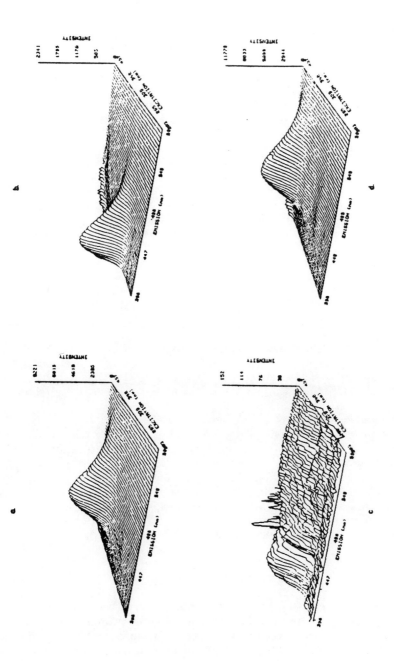

Figure 1-12. Typical EEMs of pigments eluting at: (a) 389 and 416 s, (b) 352 s, (c) 384 s, and (d) 480-s retention times. (Reprinted with permission from Clinical Chemistry.)

Because cytochromes are respiratory pigments found in all aerobic bacteria [82], the number of bacteria identifiable by primary methods may be expanded considerably.

Regardless of the specific promise of fluorescence analyses based on extracted cytochromes, it is evident that the future of primary fluorescent methods lies in using sophisticated instruments, such as the video fluorometer. The use of such instruments may allow workers to differentiate between similar bacteria based on subtle differences between the spectral properties of low-concentration pigments. Furthermore, a systematic search must be made for new pigments suitable for use in identification processes.

One additional limitation of primary methods also should be mentioned. In an analysis such as that reported by Shelly et al. [75, 76], there is no means for determining whether the fingerprint mixture of fluorophores has been generated by only one type of bacteria or by a mixture of bacteria. Hence, the assumption that an unknown sample contains only one type of bacteria must be made. Primary methods, therefore, are not likely to be useful for the analysis of mixtures of bacteria. This is not a severe constraint considering that this is a problem with almost all bacterial identification procedures.

IV. SECONDARY FLUORESCENCE METHODS

Secondary methods have been applied much more extensively in bacterial identification. One obvious reason for this is that secondary methods can be applied, in principle, to all bacteria. Also, secondary methods have been found to be more consistent and specific than primary methods.

A number of different approaches to bacterial identification can be classified as secondary methods. These approaches range in complexity from staining with fluorescent dyes to aid in morphological identification, to highly specific immunofluorescence assays (IFA). A representative sampling of recently reported applications of secondary fluorescence methods of bacterial identification is given in Table 1-4. Note that the variety of bacteria in Table 1-4 far surpasses that of Table 1-1. It is also apparent from Table 1-4 that a great majority of the applications involve IFA. Hence, IFA is the first secondary technique to be discussed in this section.

A. Immunofluorescence Assays

When introduced to the bodies of host animals, bacteria can cause an allergic response in the host. Part of this response involves development of antibodies against the antigen (the bacteria). The antibodies form complexes with the antigen in order to facilitate the removal of the antigen from the host animal's body. Identification of bacteria by IFA takes advantage of the high

Table 1-4. Recent Applications of Secondary Fluorescence Methods

Bacterium	Fluorescent tag	References
Direct IFA		
Bacterioides asaccharolyticus,	Fluorescein	83, 84
B. melaninogenicus		
subsp. intermedius		
subsp. melaninogenicus		
subsp. levii		
B. fragiles	Fluorescein	84, 86
Haemophilus sommas	Fluorescein	87
Legionella micdadei	Fluorescein	89, 92
L. pneumophila	Fluorescein	89, 92
Neisseria gonorrhoeae	Fluorescein	93, 97
Rickettsiae rickettsii	Fluorescein	98
Group A Streptococci	Fluorescein	99, 100
Group B Streptococci	Fluorescein	101
Nitrosomomas europaea	Fluorescein	102
Indirect IFA		
Actinomyces viscosis	Fluorescein, Rhodamine	103
Desulfovibrio salexigens,	Fluorescein	104
D. desulfuricans,		
D. vulgaris		
Desulfotomaculum nigrificans		
Haemophilus influenze,	Fluorescein	105
H. Pleuropneumoniae		106
Legionella pneumophila,	Fluorescein	107
Methanobacterium formiciciam,	Fluorescein	108, 109
Methanobrevibactor ruminantium,		
Methanococcus vannielli,		
Methanospirillum hugatei,		
Methanosarcina barberi,		
Methanococcus mayei		
Methanoangenium marsnigri		
Straphyloccus aureus	Fluorescein	105
Streptocuccu s pneumoniae	Fluorescein	105, 110
Flow Cytometry		
Escherichia coli	Fluorescein isothiocyanate	111
Lactobacillus brevis	Propidium iodide	
Lactobacillus casei		
Bacillus subtilis,	Ethidium bromide	112
Rhizobium japonicum		

Table 1-4. Recent Applications of Secondary Fluorescence Methods

Bacterium	Fluorescent tag	References
Total Luminescence Spectroscopy		
Scarcina lutea	Acridine orange	113
Escherichia coli	Acridine orange/	113, 115
Enterobacter cloacae	Pyrene butyric acid/	114, 115
Klebsiella pneumoniae,	Fluorescein isothiocyanate	
Proteus vulgaris	Acridine	
Shigella sonnei,		
Salmonella typhimurium,		
Streptococcus pyogenes,		
Staphylococcus aureus,		
Staphylococcus epidermis,		

degree of specificity inherent in the antibody-antigen interaction. Generally, a fluorescently tagged antibody is used to search for a specific bacterium. There are two broad classifications of IFA procedures, ie, direct and indirect. Both types of analyses can be accomplished via heterogeneous or homogenous procedures [22,23,116].

In direct IFA procedures, an antibody against the bacterium of interest is tagged with a fluorophore, usually fluorescein. Cultures or tissue sections containing the bacterium of interest then can be identified by their ability to retain the tagged antibody. The general reaction may be outlined as

$$A + Ab_f \rightarrow A\text{-}Ab_f \tag{1-17}$$

where A is the antigen and Ab_f is the tagged antibody. It is the A-Ab_f complex that must be detected to verify that a reaction has occurred. Because Ab_f and A-Ab_f are both tagged with the same fluorophore some method of distinguishing between the fluorescence of the two species must be employed. Homogeneous methods depend on a change in the character of the fluorescence of the tag after complexation has occurred so that A-Ab_f may be determined in the presence of Ab_f. Although homogenous methods are not employed commonly, it can be seen that the problem of detecting A-Ab_f in the presence of Ab_f is essentially a fluorescent mixture analysis. Hence, the application of instruments, such as the VF, designed specifically for mixture analysis is likely to increase the accuracy of homogeneous methods. To avoid the mixture analysis problem, most workers have used the more time-consuming heterogeneous methods. Such methods require that Ab_f and A-Ab_f be separated physically before analysis.

Direct IFA procedures are highly specific. Often the bacterium of interest can be detected in the presence of many similar bacteria. This is in marked contrast to primary fluorescence methods, which may fail when the unknown sample contains a mixture of fluorescent bacteria. However, the high selectivity can also be a disadvantage in that generally only a single bacterium can be detected per analysis. Hence, if a bacterial culture is known to contain a certain genus of bacteria but the exact species is unknown it may be necessary to test the culture against many different antibodies before an identification can be made. If no a priori information about the genus of the bacterium is available the problem is even more severe. For this reason, direct IFA rarely is used to identify a complete unknown. Instead, a sample is usually screened to determine whether it contains a certain bacterium. If the test result is negative, no information about the identity of the bacterium is provided. One way of partially circumventing this problem is by use of a polyvalent antibody pool. Mouton et al used this approach in their investigation of Bacteroides asaccharolyticus and Bacteroides melaninogenicus group bacteria [83]. A positive reaction would indicate that some bacterium of this group was present but it would not yield the specific identity of the antigen because each of the antibodies in the pool was tagged with identical fluorophores. A disadvantage of the direct polyvalent antibody pool technique is that each antibody must be tagged individually. Therefore time-consuming procedures are necessary to maintain a supply of reagents for routine use.

Indirect IFA requires the development of two sets of antibodies. One antibody is specific against the bacterium of interest and is not tagged. The second antibody is specific against the first antibody and is tagged. The two-step indirect IFA reaction sequence is outlined below.

$$A + Ab' \quad \rightarrow \quad A\text{-}Ab' \tag{1-18}$$

$$A\text{-}Ab' + Ab_f'' \quad \rightarrow \quad A\text{-}Ab'\text{-}Ab_f'' \tag{1-19}$$

where Ab' is the antibody against A, and Ab_f'' is the fluorescent tagged antibody against Ab'. The complex $A\text{-}Ab'\text{-}Ab_f''$ is detected in the case of a positive reaction. Both homogeneous and heterogeneous analysis schemes are possible. Heterogeneous procedures are the most commonly employed.

Although the indirect assay is generally more time consuming than direct IFA, the indirect method does offer certain advantages. Indirect methods have been used to detect Ab' in human and animal body fluids, thereby proving that the subject had at one time contacted a specific bacterium. Anderson et al searched for antibodies against Babesia gibsoni in dog serum using this method [117]. Similar studies have been conducted by a number of other workers [118-121]. It is also interesting to note that indirect IFA has an advantage over direct IFA in the use of polyvalent antisera because not every monospecific antibody in the pool must be labeled. Only the antibody against the pooled antibodies must be tagged. Hence, reagent preparation time is reduced greatly as compared to direct polyvalent antibody procedures.

In general, detection of the tagged antibodies is accomplished via fluorescence microscopy; hence, little spectral information about the tag is obtained. Future progress in IFA is dependent on the successful application of more sophisticated instrumentation. One exciting possibility for future work is the simultaneous determination of mixtures of bacteria. This might be accomplished by having a mixture of monospecific antibodies with unique fluorescent tags in a polyvalent pool. Then, positive reactions of each antibody type could be detected simultaneously by applying the principles of fluorescent mixture analysis. Because a majority of current IFA applications use fluorescein tags, it first will be necessary to develop a stock of new tags with unique spectral properties. The extensive use of fluorescein tags dates back to the earliest applications of IFA and results largely from its easily visualized fluorescence and water solubility [122,123]. In spite of the present lack of variety in fluorescent tags, preliminary work has been reported that demonstrates the applicability of IFA to multibacterial determinations [103]. Similar studies have been reported in virology [124]. Preliminary multibacteria determination was limited to binary mixtures of bacteria because the instrumentation used in the study was not sophisticated enough to accommodate the analysis of more complex mixtures. The application of more sophisticated fluorometers may provide the impetus needed to stimulate more active research in this area.

One possible scenario for the simultaneous detection of multiple bacteria via IFA/VF is outlined here. A mixture of uniquely tagged antibodies is reacted with a mixture of bacteria. The number of uniquely tagged antibodies may be quite large because the high selectivity inherent in TLS enables one to distinguish between tags with subtle spectral differences. An EEM of the mixture of tags resulting from positive reactions is acquired. Sophisticated data analysis procedures are used to determine the number of bacteria in the mixture and their identities. Because the number of bacteria that can be determined in a single sample may be quite large, the problems associated with identification of a complete unknown using IFA are reduced in this hypothetical experiment. Hence, IFA eventually may be transformed from a screening technique to a true identification process. Although an experiment such as this seems to be very speculative, such procedures may be investigated in the very near future. It is proposed that IFA and TLS are two independent, mature techniques that need only to be merged in order to promote exciting advancements such as those proposed here.

B. Flow Cytometry

The application of flow cytometry (FC) to microbiology is a fairly recent development, with early investigations limited primarily to yeast cells [125,126]. In contrast to IFA, which uses one highly specific reaction for identification of a bacterium, FC uses a combination of much less specific reactions to form a characteristic "response pattern" for each bacterium. In FC a sample of bacteria generally is stained with one or more fluorescent dyes. A list of some commonly used dyes and their cellular specificities is given in

Table 1-5 [127-129]. Fluorescent antibodies also have been used for staining DNA in some FC procedures [130]. After the bacteria are stained, the pattern of dye uptake for each bacterium may be determined using a flow cytometer. This pattern depends on the affinity of each dye for the bacterium.

The instrumentation used in FC is more sophisticated than IFA instrumentation [131]. Suspensions of stained cells flow through an examination area in a laminar sheath. Each cell individually crosses one or more laser beams to excite the adsorbed dyes. Emission from the dyes may be passed through a filter and then detected with a photomultiplier tube. A histogram of number of cells versus fluorescence intensity is constructed for each dye. These histograms may be useful fingerprinting tools. A hypothetical example of how such histograms may be used to differentiate between two bacteria is provided in Figure 1-13.

Pau et al stained Rhizobinium japonicum and Bacillus subtilis with the DNA-specific dye, acridine [112]. Differences in the FC histograms enabled these workers to separate a binary mixture of these bacteria into fractions containing only one pure bacterium. In a similar study, Hutter et al determined DNA and protein staining histograms for Escherechia coli and two species

Table 1-5.

Nucleic acid selective dyes

1. 3,8-diamino-5-diethylaminopropyl-6-phenylphenanthridium iodide (PI)
2. 4',6-diamidino-2-phenylindole (DAPI)
3. 3,6-diamino-10-methyl-acridiniumchloride (AF)
4. 7-amino-3-(β-D-ribofuranosyl)pyrazole-(4,3-d)pyrimidine (Formycil)
5. Acridine (AC)
6. 3,6-dimethylamino acridine (AO)

Protein selective dyes

1. Fluorescein isothiocyanate (FITC)
2. 4-phenylspiro[furan-2(3),1'-phthalan]-3,3'-dione (FC)
3. 5-dimethylamino-1-naphthalenesulfonylchloride (DANS)
4. Tetramethylrhodamine isothiocyanate (TRITC)
5. Sulforhodamine 101 (SR101)
6. Chloro-4-nitrolanzo-2-oxa-1,3-diazole (NBD)
7. 4-acetamido-4'-isothiocyanatostilbene-2,2'-disulfonic acid (SITS)

Lipid selective dyes

1. 4,4'-bis[4-(3-sulfanilo)-6-[bis-2-hydroxyethylamino]-1,3,5-[triazen-2] aminostilbene-2,2'-disulfonic acid
2. Pyrene butyric acid (PBA)

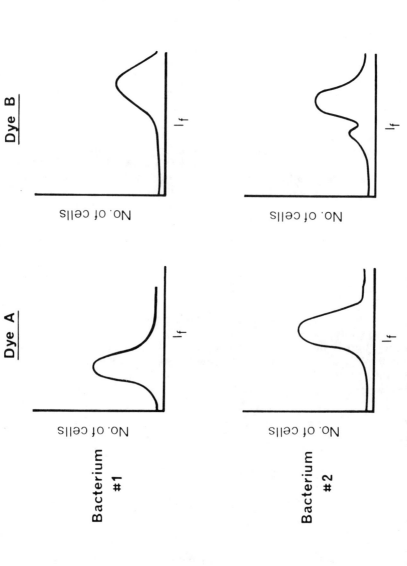

Figure 1-13. Hypothetical flow cytograms illustrating differential adsorption patterns of dyes A and B for bacteria 1 and 2.

of Lactobacillus [111]. Differentiation of these bacteria was possible based on their DNA staining profiles.

Instrumental limitations in FC restrict users to a maximum of three dyes during a given analysis [128]. Commonly only one or two dyes are used at one time. Hence, the number of parameters contributing to the FC fingerprint is inherently small. This imposes a limit to the number of bacteria that can be differentiated or identified under a given set of experimental conditions. In order to increase the uniqueness of the FC response patterns it is desirable to increase the number of dyes used. This problem may be viewed as the basic problem of analysis of a mixture of fluorophores. As such, it seems likely that the application of TLS to a FC-type fingerprinting technique may provide a useful identification tool. It is this reasoning that has formed the basis for the development of mixed-dye fluorometry.

C. Mixed-Dye Fluorometry

Total luminescence spectroscopy was first applied to the identification of bacteria by Ginell and Feuchtbaum [113]. In their original study, EEM-type data were recorded for both primary and secondary fluorophores associated with E. coli and Sarcina lutea. Both bacteria exhibited primary fluorescence. However, the characteristics of this fluorescence were found to be indistinct and not useful for fingerprinting. After both bacteria were stained with acridine orange (3,6-dimethylamino acridine) differences in the EEMs of the bacteria were observed. These spectral differences probably resulted from differences in the amount of dye adsorbed by the bacteria; as well, spectral shifts in the dye resulted from adsorption site characteristics. Ginnel and Feuchtbaum used a conventional fluorometer to acquire the data. The tedious data acquisition process and the low-resolution nature of their hand-plotted EEMs failed to stimulate any immediate followup studies. Further work using TLS in secondary fluorescence identification methods was not reported until Shelly et al applied the VF to the problem [114,115]. The technique developed by Shelly et al is named mixed-dye fluorometry (MDF).

In MDF a mixture of dyes is used to stain the bacteria. This is similar to the basis of identification of bacteria by FC. In fact, MDF utilizes the same principle of selective dye adsorption as does FC to identify bacteria. However, in MDF the number of dyes in the mixture and, hence, the number of parameters contributing to the fingerprint is less restricted than in FC. Shelly et al used a combination of four dyes to form a fingerprint response pattern. These dyes were fluorescein isothiocyanate, pyrene butyric acid, acridine, and acridine orange. Information obtained from the literature and provided in Table 1-5 verifies that this mixture is capable of staining the major cellular components common to all bacteria. The details of these original MDF studies are discussed here.

Although four dyes were used for staining, the dyes were prepared in two binary mixtures. Staining of each sample of bacteria with only one of these mixtures at a time allowed the workers to reduce the complexity of data interpretation by avoiding a high degree of spectral overlap between the dyes in any given mixture. Two distinct approaches were explored to determine quantitatively the amount of each dye adsorbed in the staining process. One approach was a direct process in which the bacteria would be examined directly once they were stained. The other, indirect approach did not require direct examination of the stained bacteria.

The details of these methods can perhaps be best understood by a consideration of Figure 1-14. In this figure, a hypothetical EEM of a two-component dye mixture is plotted. This EEM is assigned the name Q_1 for the purposes of the present discussion. Also shown in Figure 1-14 are the dye adsorption EEMs Q_{2A} and Q_{2B}, and the residual dye EEMs Q_{3A} and Q_{3B} for two different bacteria, A and B. A direct evaluation of the dye adsorption properties of bacteria A and B would involve direct acquisition of Q_{2A} and Q_{2B}. These EEMs can be obtained directly only by an examination of the stained cells. Shelly et al tested the direct method and found that much of the spectral information pertaining to the adsorbed dyes was obscured by high scattered light levels. Such scattered light problems are inherent in video fluoremetric evaluation of highly turbid samples, such as cell suspensions. As a means of avoiding a direct examination of the cell suspensions, Shelly et al developed an indirect method of acquiring spectra Q_{2A} and Q_{2B}. In this method spectra Q_1, Q_{3A}, and Q_{3B} are acquired directly. The spectra Q_{3A} and Q_{3B} do not suffer from as large an interference from scattered light because the bacteria have been centrifuged out of these dye mixtures prior to analysis. The EEMs Q_{2A} and Q_{2B} were then calculated using the equations:

$$Q_{2A} = Q_1 - Q_{3A} \tag{1-20}$$

and

$$Q_{2B} = Q_1 - Q_{3B} \tag{1-21}$$

Although this may seem to be the last necessary step in data reduction for the indirect method, such is not the case. Shelly et al found that minor differences between Q_{2A} and Q_{2B} could not be observed easily. Hence, as a method of accentuating these differences they defined two new matrices that could be obtained by using the following equations:

$$R_1 = Q_{2A}/Q_1 \tag{1-18}$$

and

$$R_2 = Q_{2B}/Q_1 \tag{1-19}$$

Using RMR data reduction methods, it is possible to determine from R_1 and R_2 the relative concentration of each dye adsorbed by the bacteria. These

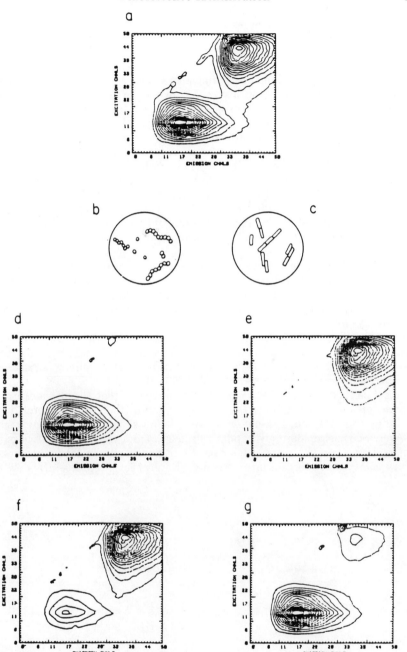

Figure 1-14. Contour plots of hypothetical dye mixtures illustrating the MDF procedure, where: a is matrix Q_1, b is bacterium A, c is bacterium B, d is matrix Q_{2A}, e is matrix Q_{2B}, f is matrix Q_{3A}, g is matrix Q_{3B}. (Reprinted with permission from Clinical Chemistry.)

ratio values tend to emphasize minor differences in the adsorption EEMs; hence, they are a more convenient fingerprinting tool than are the adsorption EEMs.

The application of the above principles to real bacteria may be explained using a few examples. Using bacteria grown in brain-heart infusion media, Shelly et al used two dye mixtures to identify each bacterium. These mixtures are labeled arbitrarily Y and Z. The composition of these mixtures is given in Table 1-6. Because two mixtures of dyes were used in each case, two dye adsorption and two ratio matrices were evaluated for each bacterium. As a means of comparing the selective dye adsorption properties of the bacteria tested, the ratio values for each dye and each bacterium are plotted in bar graph format in Figure 1-15A, B. The uniqueness of these bar graphs indicates that MDF does provide useful fingerprinting information for the nine bacteria tested. In replicate experiments with new cell cultures these ratio data were found to be highly reproducible.

It is also desirable to have some means of expressing the degree of similarity or uniqueness of the bar graphs of Figure 1-15 A, B. Shelly et al used the method of double ratio plots to perform cluster analysis-type pattern recognition studies. A double ratio plot is constructed by plotting the ratios of the dye response ratios of two dyes on each axis of a two-axis system. When four dyes are used, three distinct double ratio graphs can be defined. The results of plotting the bar graph data of Figure 1-15A, B on the three possible double ratio systems are shown in Figure 16. Shelly et al made the following observations concerning the patterns in these plots. First, the general lack of overlap between bacteria indicates that the dye adsorption characteristics of each bacterium are unique and, therefore, a useful finger-printing tool. Second, the gram-positive organisms Staphylococcus aureus and Staphylococcus epidermidis tend to lie away from the gram-negative organisms. This is probably because of increased peptidoglycan layer thickness in the gram-positive organisms.   Finally, Enterobacter cloacae, E. coli and

Table 1-6.

Dye Mixture	Components[a]
Y	8.5×10^{-4} mg/mL PBA 4.3×10^{-4} mg/mL AO
Z	8.3×10^{-4} mg/mL AC 4.0×10^{5} mg/mL FITC

[a] for definitions of component abbreviations see Table 1-5.

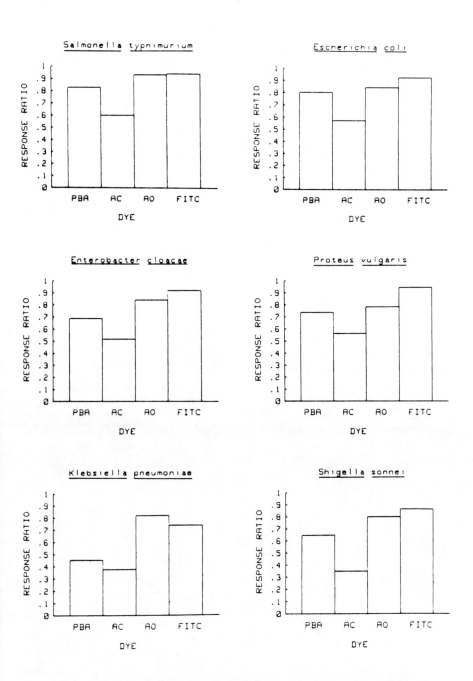

Figure 1-15A. Bar graph representation of dye adsorption patterns for gram-negative bacteria. (Reprinted with permission from Clinical Chemistry.)

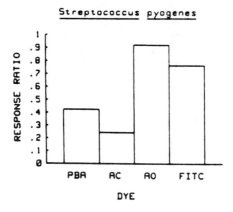

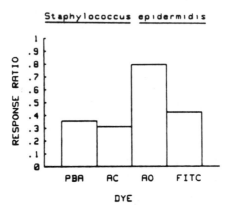

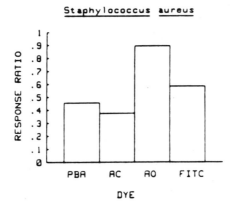

Figure 1-15B. Bar graph representation of dye adsorption patterns for gram-positive bacteria. (Reprinted with permission from Clinical Chemistry.)

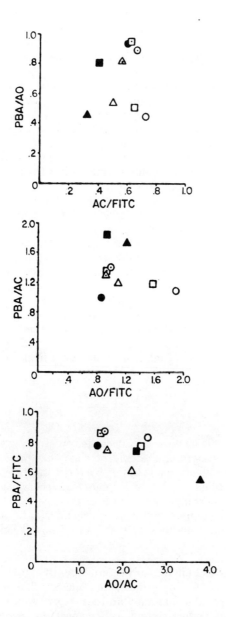

Figure 1-16.   Double ratio plots illustrating the differences in dye adsorption patterns for all bacteria tested. The organisms have been plotted according to the code scheme: Klebsiella pneumoniae (△), Staph. epidermidis (O), Staph. aureus (□), Strep. pyogenes (▲), Proteus vulgaris (●), Shigella sonnei (■), Ent. cloacae (⬚), Sal. typhimurium (Ө), and E. coli (⊡). (Reprinted with permission from Clinical Chemistry.)

Salmonella typhinurium all group very closely to each other. This indicates that this particular combination of dyes may not be optimum for differentiating these three bacteria. Other, more complicated mixtures of dyes may be required to form good fingerprints of these three bacteria.

The preliminary study described above is not a conclusive case in support of the utility of MDF as a fingerprinting tool. However, certain general conclusions may be drawn. First, the degree of success of this initial work warrants the continuance of work in identification of bacteria by MDF. The method is simple and rapid (typical analysis time, 2 h) and highly reproducible. Future work in this area should include an evaluation of new dye mixtures that may provide more unique fingerprints than did the dyes used in the initial study. Also, more sophisticated analysis techniques should be applied to the data. It is possible that the application of Fourier transform-based pattern recognition techniques to the Q_3 matrices will bypass the necessity of ratio calculations and human interpretation of double ratio plots. Finally, detailed studies of the factors that influence a given bacterium's affinity to a given dye should be conducted. Information produced from such a study may be useful in determining what properties are desirable in new dyes to be tested in the MDF procedure. As the technique is further evaluated, more bacteria must be tested. Although MDF is far from being a clinically accepted technique at present, it does seem to have good potential for eventual clinical use. One drawback of the method is that it is limited fundamentally to use as a means of evaluating pure cultures with only one bacterium present. However, this is a limitation of many bacterial identification schemes.

V. CONCLUSION

It is hoped that this chapter has succeeded in stressing several major concepts. From the introduction, it should be clear that there is a need for the development of instrumental methods of bacterial identification. In the past, this need has been filled by the development of techniques that attempt to identify bacteria by decomposing them into their chemical components and then analyzing these complex mixtures of chemicals. In general, this is not true of the fluorescence methods of analysis that have been discussed in this chapter.

The fluorescence techniques for bacterial identification have been divided into two major categories. The primary methods are limited by virtue of the relative scarcity of suitable natural fluorescent pigments in bacteria. It was shown that by applying total luminescence spectroscopy techniques to the analysis of mixtures of pigments the information yielded in primary fluorescence studies was greatly increased. This introduced an underlying theme that reappeared throughout the chapter. That is, the use of fluorescence spectroscopy in bacterial identification can be greatly aided by application of the principles of fluorescence mixture analysis as developed in Section II.

This theme was recurrent in the discussion of secondary methods. The use of mixture analysis was proposed as a means of greatly expanding the capabilities of IFA for true unknown identification and for the simultaneous analysis of mixtures of bacteria. The flow cytometry method also can be viewed as a means of mixture analysis. The limitations of FC severely restrict the number of dyes that may be used to fingerprint bacteria. By application of more advanced mixture analysis techniques, a multiple-dye fingerprinting technique was developed that was not as restricted as FC. These three secondary methods are related to one another by virtue of their dependence on mixture analysis techniques for future improvements. However, they are very different in principle. The technique of IFA is the only technique that relies on a single specific reaction for identification of a bacterium. The other two techniques, FC and MDF, use a combination of less specific reactions to form characteristic response patterns for various bacteria.

Other approaches to the identification of bacteria by fluorescence spectroscopy also may be fruitful in the future. For example, fluorescence techniques may be used to characterize bacteria based on the unique mixture of enzymes inherent in each bacteria type. Enzyme content may be monitored by IFA, for example [132]. Based on such possibilities, a technique may be envisioned wherein a mixture of uniquely tagged antibodies against a broad set of enzymes is used to fingerprint a bacterium by its enzyme content. Several researchers proposed the use of fluorescence evaluation of enzyme patterns to identify bacteria [133,134]. These workers bound fluorophores to various substrates by linkages that could be cleaved only by specific enzymes. After enzymatic cleavage the released fluorophores were detected and found to form patterns characteristic of the bacteria from which the enzymes were released. Instrumentation in these studies was highly specialized and the techniques never gained widespread clinical acceptance.

Again, these new prospects of bacterial identification require a good technique for the analysis of mixtures of fluorophores. From the preceeding discussion and from the other sections of this chapter, it may be concluded that many of the exciting new advances in fluorescence-based identification of bacteria stem from a merger of the techniques of TLS with previously existing microbiological methods. This merger between two seemingly independent fields already has produced some interesting and exciting results and holds the promise of many new applications in bacterial identification.

ACKNOWLEDGMENT

The authors gratefully acknowledge support of the NIH (Grant AI19916) during the preparation of this manuscript.

VI. REFERENCES

1. Irwin, W. J. "Analytical Pyrolysis: A Comprehensive Guide". Marcel Dekker, Inc.: New York, 1982; Chap. 8, pp. 381-431.
2. Oyama, V. I. Nature (London), 1963, 200, 1058.
3. Simmonds, P. G. Appl. Microbiol., 1970 20, 567.
4. Reiner, E. Nature (London), 1965, 206, 1272.
5. Hus In't Veld, J. H. J.; Meuzelaar, H. L. C.; Tom, A. Appl. Microbiol., 1973, 26, 92.
6. Stack, M. V.; Donohue, H. D.; Tyler, J. E. Appl. Environ. Microbiol., 1978, 35, 45.
7. Reiner, E.; Beam, R. E.; Kubica, G. P. Am. Rev. Respir. Dis., 1969, 99, 750.
8. Reiner, E.; Kubica, G. P. Am. Rev. Respir. Dis., 1969, 99, 42.
9. Reiner, E.; Moran, T. F. Adv. Chem. Series, 1983, 203, 705.
10. Menger, F. E.; Epstein, G. A.; Goldberg, D. A.; Reiner, E. Anal. Chem., 1972, 44, 423.
11. French, G. L.; Gutteridge, C. S.; Phillips, I. J. Appl. Bacteriol., 1980, 49, 505.
12. Milana, R.; Dimov, N.; Dimitrova, M. Chromatographia, 1983, 17, 29.
13. Brooks, J. B.; Moss, C. W.; Dowell, V. R. J. Bacteriol., 1969, 100, 528.
14. Thadepalli, H.; Congopadhyay, P. K.; Ansari, A.; Overturf, G. D.; Dahwan, V. K.; Mandal, A. K. J. Clin. Invest., 1982, 69, 979.
15. Hayward, N. J. J. Chromatogr., 1983, 274, 27.
16. Henis, Y.; Gould, J. R.; Alexander, M. Appl. Microbiol., 1966, 14, 513.
17. Thomas, L. C.; Greenstreet, J. E. S. Spectrochim. Acta, 1954, 6, 302.
18. Riddle, J. W.; Kabler, P. N.; Kenner, B. A.; Bordner, R. H.; Rockwood, S. A.; Stevenson, H. J. R. J. Bacteriol., 1956, 72, 593.
19. Rideal, E. K.; Adams, D. M. Chem. Ind. (London), 1957, 35, 762.
20. Booth, G. H.; Miller, J. D. A.; Paisley, H. M.; Saleh, A. M. J. Gen. Microbiol., 1966, 44, 83.
21. Slots, J.; Reynolds, H. S. J. Clin. Microbiol., 1982, 16(6), 1148.
22. Maggio, E. T. In "Immunoassays: Clinical and Laboratory Techniques for the 1980's"; Alan R. Liss, Inc.: New York, 1980; pp. 1-12.
23. Ullman, E. F.; Bellet, N. F.; Brinkley, J. M.; Zuk, R. F. In "Immuno-assays: Clinical and Laboratory Techniques for the 1980's"; Alan R. Liss, Inc.: New York, 1980; pp. 13-43.
24. Parker, C. A. "Photo Luminescence of Solutions"; Elsevier Publishing Co.: New York, 1968; Chap. 1, pp. 1-22.
25. Kasha, M. In "Fluorescence Theory, Instrumentation and Practice"; Guilbault, G. G., ed.; Marcel Dekker, Inc.: New York, 1967; Chap. 4, pp. 201-232.
26. Becker, R. S. "Theory and Interpretation of Fluorescence and Phosphorescence Analysis"; Wiley Interscience: New York, 1969.
27. Hercules, D. M. In "Fluorescence and Phosphorescence Analysis"; Hercules, D. M., ed.; Wiley Interscience: New York, 1966; Chap. 1, pp. 1-40.

28. Wright, J. C.; Wirth, M. J. Anal. Chem., 1980, 52, 988A.
29. Wright, J. C.; Wirth, M. J. Anal. Chem., 1980, 52, 1087A.
30. Warner, I. M.; McGowen, L. B. CRC Crit. Rev. Anal. Chem., 1982, 13, 155.
31. Love, L. J. C.; Shaver, L. A. Anal. Chem., 1976, 48, 364A.
32. Knorr, F. J.; Harris, J. M. Anal. Chem., 1981, 53, 272.
33. Giering, L. P. Ind. Res., 1978, 20, 134.
34. Haugen, G. H.; Raby, G. A.; Rigdon, L. P. Chem. Inst., 1975, 6, 205.
35. Bentz, A. P. Anal. Chem., 1976, 48, 455A.
36. Gaugh, T. H. Science, 1979, 203, 1330.
37. Warner, I. M.; Callis, J. B.; Davidson, E. R.; Christian, G. D. Clin. Chem., 1976, 22, 1483.
38. Johnson, D. W.; Gladden, J. A.; Callis, J. B.; Christian, G. D. Rev. Sci. Instrum., 1978, 49, 54.
39. Warner, I. M.; Fogarty, M. P.; Shelly, D. C. Anal. Chim. Acta, 1979, 109, 361.
40. Warner, I. M. In "Contemporary Topics in Analytical and Clinical Chemistry", Vol. 4; Hercules, D. M.; Hieftje, G. M.; Snyder, L. R.; Evenson, M. A., eds.; Plenum Publishing Corp.: New York, 1982; Chap. 5.
41. Talmi, Y. Anal. Chem., 1975, 47, 658A.
42. Talmi, Y. Anal. Chem., 1975, 47, 699A.
43. Talmi, Y.; Baker, D. C.; Jadamec, J. R.; Saner, W. A. Anal. Chem., 1978, 50, 936A.
44. Wong, M.; Oldham, P.; Ho, C-N.; Warner, I. M. Chem., Biomed. Environ. Instrum., 1982, 12(3), 185.
45. Fogarty, M. P.; Warner, I. M. Appl. Spectrosc., 1982, 36, 460.
46. Fogarty, M. P.; Warner, I. M. Anal. Chem., 1981, 53, 259.
47. Fogarty, M. P.; Shelly, D. C.; Warner, I. M. HRC & CC, 1981, 4, 561.
48. Shelly, D. C.; Fogarty, M. P.; Warner, I. M., HRC & CC, 1981, 4, 616.
49. Hershberger, L. W.; Callis, J. B.; Christian, G. D. Anal. Chem., 1981, 53, 971.
50. Shelly, D. C. "Two-Dimensional Fluorescence Fingerprinting Using a Rapid Scanning Fluorometer"; PhD Dissertation, Texas A&M University, College Station, Texas 1982.
51. Rossi, T. M.; Quarles, J. M.; Warner, I. M. Anal. Lett., 1982, 15(B13), 1083.
52. Warner, I. M.; Christian, G. D.; Davidson, E. R.; Callis, J. B. Anal. Chem., 1977, 49, 564.
53. Warner, I. M.; Davidson, E. R.; Christian, G. D. Anal. Chem., 1977, 49, 2155.
54. Ho, C-N.; Christian, G. D.; Davidson, E. R. Anal. Chem., 1978, 50, 1100.
55. Ho, C-N.; Christian, G. D.; Davidson, E. R. Anal. Chem., 1980, 52, 1071.
56. Rossi, T. M.; Warner, I. M. Appl. Spectrosc., 1983.
57. Rossi, T. M.; Warner, I. M. Appl. Spectrosc., 1985.
58. Fogarty, M. P.; Warner, I. M. Appl. Spectrosc., 1980, 34, 438.
59. Bracewell, R. N. "The Fourier Transform and Its Applications", The McGraw Hill Book Co.: New York, 1978.

60. Gonzalez, R. C.; Wintz, P. "Digital Image Processing", Addison-Wesley Publishing Co.: Reading, MA, 1977.
61. Isenhour, T. L.; Lam, R. B. Anal. Chem., 1981, 53, 1179.
62. Lam, R. B.; Isenhour, T. L. Anal. Chem., 1982, 54, 154.
63. Horlick, G.; Hieftje, G. M. In "Contemporary Topics in Analytical Chemistry"; Hercules, D. M., ed.; Plenum Press: New York, 1978; Chap. 4, pp. 153-216.
64. Ng, R. C. L.; Horlick, G. Spectrochim. Acta, 1981, 36B, 529.
65. Harding, G. K. M.; Sutter, V. L.; Finegold, S. M.; Bricknell, K. S. J. Clin. Microbiol., 1976, 4, 354.
66. Ellner, P. D.; Granato, P. A.; May, C. B. Appl. Microbiol., 1973, 26, 904.
67. Melo, T. B.; Johnsson, M. Dermatologica, 1982, 164, 167.
68. Hulcher, F. H. Biochemistry, 1982, 21, 4491.
69. Wong, W. C.; Fletcher, J. T.; Unsworth, B. A.; Preece, T. F. J. Appl. Bacteriol., 1982, 52, 43.
70. Shaw, B. G.; Latty, J. B. J. Appl. Bacteriol., 1982, 52, 219.
71. Gilardi, G. L. Appl. Microbiol., 1971, 21, 414.
72. Gilardi, G. L. Am. J. Med. Tech., 1972, 38, 65.
73. Suzuki, A.; Miyagawa, T.; Goto, M. Bull. Chem. Soc. Jpn., 1972, 45, 2198.
74. Suzuki, A.; Goto, M. Bull. Chem. Soc. Jpn., 1971, 44, 1869.
75. Shelly, D. C.; Warner, I. M.; Quarles, J. M. Clin. Chem., 1980, 26, 1127.
76. Shelly, D. C.; Quarles, J. M.; Warner, I. M. Clin. Chem., 1980, 26, 1419.
77. Shuvalov, V. A.; Parson, W. W. Biochim. Biophys. Acta, 1981, 638, 50.
78. Kaiser, G. H.; Beck, J.; Von Schutz, J. U.; Wolf, H. C. Biochem. Biophys. Acta, 1981, 634, 153.
79. Chow, A. W.; Patton, V.; Greze, L. B. J. Clin. Microbiol., 1975, 2, 546.
80. King, E. O.; Ward, M. K.; Raney, D. E. J. Lab. Clin. Med., 1954, 44, 301.
81. Wood, P. M. Anal. Biochem., 1981, 111, 235.
82. Forbisher, M. In "Fundamentals of Microbiology"; W. B. Saunders Co.: Philadelphia, 1957; Chap. 29, p. 385.
83. Mouton, C.; Hammond, P.; Slots, J.; Genco, R. J. J. Clin. Microbiol., 1980, 11, 682.
84. Weissfeld, A. S.; Sonnenwirth, A. C. J. Clin. Microbiol., 1981, 13, 798.
85. Grzelak-Puczynska, I.; Masel-Mikolajczyk, F. J. Appl. Bacteriol., 1981, 51, 217.
86. De-Girolami, P. C.; Mepani, C. P. Am. J. Clin. Pathol., 1981, 76, 78.
87. Humphrey, J. D.; Little, P. B.; Stephens, L. R.; Barnum, D. A.; Doig, P. A.; Thorson, J. Am. J. Vet. Res., 1982, 43, 791.
88. Cordes, L. G.; Myerowitz, R. L.; Pasculle, A. W.; Corcoran, L.; Thompson, T. A.; Gorman, G. W.; Patton, C. M. J. Clin. Microbiol., 1981, 13, 720.
89. Lowry, B.S.; Vega, Jr., F. G.; Hedlund, K. W. Am. J. Clin. Pathol., 1982, 77, 601.

90. Winn, Jr., W. C.; Glavin, F. L.; Perl, D. P.; Keller, J. L.; Andres, T. L.; Brown, T. M.; Coffin, C. M.; Sensecqua, J. E.; Roman, L. N.; Craighead, J. E. Arch. Pathol. Lab. Med., 1978, 102, 344.
91. Gerber, J. E. ; Casey, C. A.; Martin, P.; Winn, Jr., W. C. Am. J. Clin. Pathol., 1981, 76, 816.
92. Winn, Jr., W. C.; Cherry, W. B.; Frank, R. O.; Casey, C. A.; Broome, C. V. J. Clin. Microbiol., 1980, 11, 59.
93. Young, H.; McMillan, A. Br. J. Vener. Dis., 1982, 58, 109.
94. Freundlich, L. F.; Rosenthal, S. L.; Hochberg, F. P.; Trogele, M. R. Am. J. Clin. Pathol., 1982, 77, 456.
95. Arko, R. J.; Finley-Price, K. G.; Wong, K.-H.; Johnson, S. R.; Reising, G. J. Clin. Microbiol., 1982, 15, 435.
96. Carlson, B. L.; Haley, M. S.; Kelly, J. R.; McCormack, W. M. J. Clin. Microbiol., 1982, 15, 231.
97. Johnston, N. A. Br. J. Vener. Dis., 1981, 57, 315.
98. Hebert, G. A.; Tzianabos, T.; Gamble, W. C.; Chappell, W. A. J. Clin. Microbiol., 1980, 11, 503.
99. Lauer, B. A.; Reller, L. B.; Mirrett, S. J. Clin. Microbiol., 1983, 17, 338.
100. Ederer, G. M.; Chapman, S. S. Appl. Microbiol., 1972, 24, 160.
101. Boyer, K. M.; Gadzala, C. A.; Kelly, P. C.; Burd, L. C.; Gotoff, S. P. J. Clin. Microbiol., 1981, 14, 550.
102. Belser, L. W.; Schmidt, E. L. Appl. Environ. Microbiol., 1978, 36, 589.
103. Gillis, T. P.; Thompson, J. J. J. Clin. Microbiol., 1978, 8, 351.
104. Smith, A. D. Arch. Microbiol., 1982, 133, 118.
105. Clausen, C. R. J. Clin. Microbiol., 1981, 13, 1119.
106. Rosendal, S.; Boyd, D. A. J. Clin. Microbiol., 1982, 16, 840.
107. Skinner, A. R.; Swann, A. Histochem., 1981, 71, 581.
108. Conway de Macario, E.; Macario, A. L. J.; Wolin, M. J. J. Bacteriol., 1982, 149, 320.
109. Conway de Macario, E.; Wolin, J. J.; Macario, A. J. L. J. Bacteriol., 1982, 149, 316.
110. Wicher, K.; Kalinka, C.; Mlodozeniec, P.; Rose, N. R. Am. J. Clin. Pathol., 1982, 77, 72.
111. Hutter, K.-J.; Eipel, H. E. J. Gen. Microbiol., 1979, 113, 369.
112. Pau, A. S.; Cowles, J. R.; Oro, J.; Bartel, A.; Hungerford, E. Arch. Microbiol., 1979, 120, 271.
113. Ginell, R.; Feuchtbaum, R. F. J. Appl. Bacteriol., 1971, 35, 29.
114. Shelly, D. C.; Warner, I. M.; Quarles, J. M. Clin. Chem., 1983, 29, 240.
115. Shelly, D. C.; Quarles, J. M.; Warner, I. M. Anal. Lett., 1981, 14(B13), 1111.
116. O'Donnel, C. M.; Suffin, S. C. Anal. Chem., 1979, 51, 33A.
117. Anderson, J. F.; Magnerelli, L. A.; Sulzer, A. J. Am. J. Vet. Res., 1980, 41, 2102.
118. Chisolm, E. S.; Ruebush II, T. K.; Sulzer, A. J.; Healy, G. R. Am. J. Trop. Med. Hyg., 1978, 27, 14.
119. Black, C. M.; Pine, L.; Reimer, C. B.; Benson, R. F.; Wells, T. W. J. Clin. Microbiol., 1982, 15, 1077.
120. Hanna, L.; Keshishyan, H. J. Clin. Microbiol., 1980, 12, 409.

121. Greenberg, R. N.; Sanders, C. V.; Lewis, A. C.; Marier, R. L. Am. J. Med., 1981, 71, 841.
122. Coons, A. H.; Creech, H. J.; Jones, R. N. Proc. Soc. Expl. Biol. Med., 1941, 47, 200.
123. Coons, A. H.; Creech, H. J.; Jones, R. N.; Berliner, E. J. Immunol., 1942, 45, 159.
124. Hiller, G.; Jungwirth, C.; Weber, K. Exp. Cell. Res., 1981, 132, 81.
125. Hutter, K.-J.; Stohr, M. Microbios. Lett., 1979, 10, 121.
126. Hutter, K.-J., Eipel, H. E. Eur. J. Appl. Microbiol. Biotechnol., 1979, 6, 223.
127. Stohr, M.; Vogt. Schaden, M.; Knobloch, M.; Vogel, R. Stain Technol., 1978, 53, 205.
128. Shapiro, H. M. J. Histochem. Cytochem., 1977, 25, 976.
129. Latt, S. A. In "Flow Cytometry and Sorting"; Melamed, R. R.; Mullaney, P. F.; Mendelsohn, M. L., eds.; John Wiley and Sons: New York, 1979; Chap. 15, p. 263.
130. Crissman, H. A.; Stevenson, A. P.; Orlicky, D. J.; Kissane, R. J. Stain Technol., 1978, 53, 321.
131. Horan, P. K.; Wheeless, L. L. Science, 1977, 198, 149.
132. Hargis, J. W.; Husain, S. S. Can. J. Microbiol., 1981, 27, 1076.
133. Dyer, D. L. U.S. Patent 3,551,295, Northrop Corp., Chem. Abstr. No. 74073316, 1970.
134. Yaron, A.; Carmel, A. U.S. Patent 4,314,936, Yeda Research and Development Ltd., Chem. Abstr. No. 97068544, 1982.

2. DETECTION OF MICROORGANISMS BY BIO AND CHEMILUMINESCENCE TECHNIQUES

Harold A. Neufeld, Judith G. Pace, and Richard W. Hutchinson

I. INTRODUCTION

Detection of microorganisms is complicated because of their small mass. If time permits, this is no real problem because culture techniques permit colony development; however, if there is a need for rapid detection, this becomes a real problem. Many of the current techniques depend on a specific enzyme assay, which is usually sensitive enough, but the particular enzyme may not be distributed universally in microorganisms. For example, the use of the fluorescent antibody technique both lends sensitivity in detection and has the advantage of permitting identification. There are, however, two major problems: the first is that the technique requires the preparation and storage of a large number of fluorescent antibodies, and the second is the time required for identification.

Development of luminescent techniques overcomes the problems both of speed and storage of sensitive reagents. In this chapter two approaches to detection by luminescent techniques are discussed. The first is a chemiluminescent technique that depends on the hemincatalyzed luminescence of luminol. This technique has the advantage of not depending on the presence of an enzyme. The procedure is based on the presumption that most organisms contain some form of hematin iron. The second is the luminescence of the firefly luciferin-luciferase system, which is based on the fact that all microorganisms, except for viruses, contain ATP. This is also a very sensitive method. Both methods permit the detection of 10^3 organisms per milliliter. The luminescent techniques have the disadvantage of not permitting the identification of the microorganisms.

II. BIOLUMINESCENCE

In such organisms as some ocean-dwelling bacteria, certain fungi, and some insects, oxidoreduction-type enzyme-catalyzed reactions take place in which the free-energy change excites a molecule to a high-energy state, which then is followed by the return of the excited molecule to the ground state. If, during this process, the return to the ground state is accompanied by the emission of visible light with little or no heat, the phenomenon is known as bioluminescence [1].

A. Detection of ATP Using Luciferin-Luciferase

The luminescence of the firefly is caused by a sequence of reactions involving firefly luciferin (L), ATP [Figure 2-1], luciferase (E), magnesium, and oxygen [2].

$$Mg^{+2}$$

$$LH_2 + ATP + E \rightleftharpoons E - LH_2 - AMP + PP \qquad (2\text{-}1)$$

$$E \cdot LH_2 \cdot AMP + O_2 \rightleftharpoons (E \cdot L \cdot AMP)^* + H_2O \qquad (2\text{-}2)$$

$$(E \cdot L \cdot AMP) + light$$

The reaction does not occur in the absence of oxygen and it is generally believed to be absolutely dependent on the presence of ATP. According to Seliger and McElroy [3], the quantum efficiency of the reaction is nearly 100%, with one quantum of light being released for every molecule of ATP used. The structure of firefly luciferin in the oxidized and reduced states is shown [2] in Figure 2-2.

The reaction is so specific for ATP that using a relatively crude preparation of luciferase, luciferin, and magnesium ion, a very specific and sensitive assay for ATP can be developed. This method has been published by Neufeld and others [4-14].

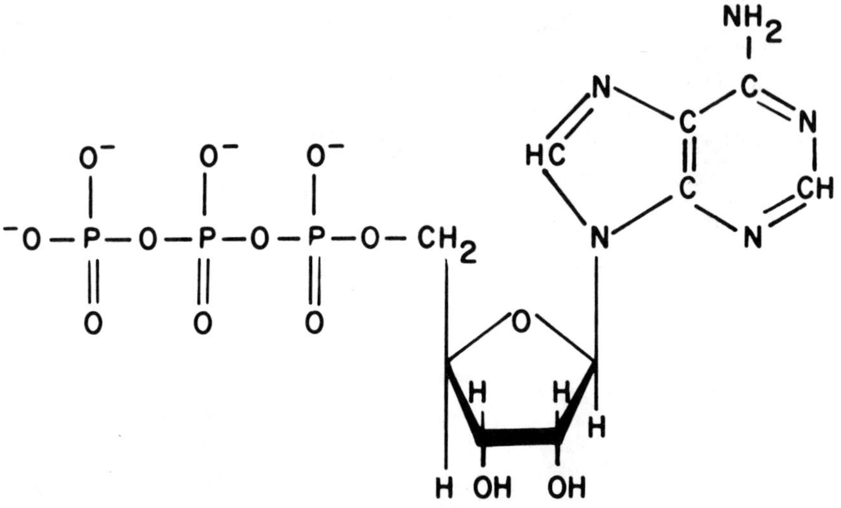

Figure 2-1. Adenosine triphosphate

In order to obtain reliable and consistent results, particular attention must be paid to the reagents used and to the phenomenon that we have called inherent light. Inherent light has been defined as that light which is present even in partially purified luciferase preparations and in the absence of exogenous ATP.

B. Standardization of the Assay

Partially purified firefly luciferase. Partially purified luciferase was prepared from a stock of desiccated and frozen firefly tails by the following procedure: 10 g of firefly tails was ground in a mortar and pestle to a fine powder. This powder was then placed into a beaker containing 50 ml of acetone that previously had been chilled to minus 20°C or below. The powder was well stirred in the acetone [2] and then the mixture was filtered through a Buchner funnel. The powder was dried by drawing air through it until the odor of acetone was absent. This acetone powder then was used immediately in the succeeding steps or stored dry in the deep freeze until required.

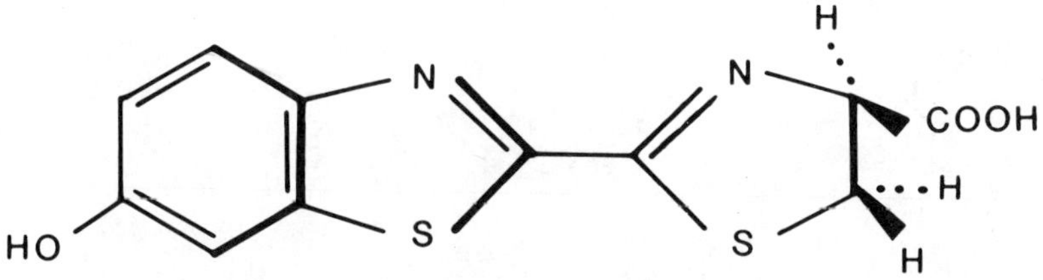

D (−) LUCIFERIN

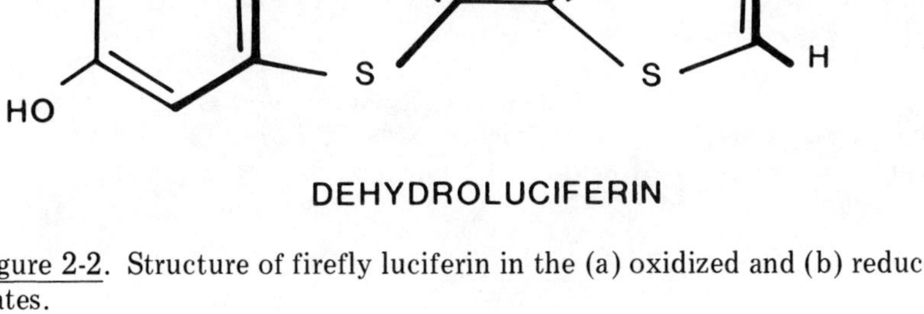

DEHYDROLUCIFERIN

Figure 2-2. Structure of firefly luciferin in the (a) oxidized and (b) reduced states.

The acetone powder was then extracted with approximately 25-35 mL of 0.1 M glycylglycine buffer, pH 7.8, which contained 1×10^{-3} M Versene by homogenization in the cold with a Potter-Elvehjem-type homogenizer. The mixture was then centrifuged in a Sorvall SS-1 at 5000 rpm for 20 min. and the supernatant fluid was saved. The precipitate was reextracted with 20 mL of 0.017 M NaOH containing 1×10^{-3} M Versene and recentrifuged. The supernatant fluid was combined with the first supernatant fluid. The NaOH extraction step was repeated one time with the supernatant fluid combined with the others to give a volume of between 50 and 75 mL. The precipitate was discarded. The combined supernatant fluid was labeled fraction I.

Fraction I was then subjected to stepwise ammonium sulfate fractionation in the following manner: solid ammonium sulfate was added to bring its concentration 20% of saturation. The resulting precipitate was dissolved in 10 mL or 0.01 glycylglycine buffer, pH 7.8, and labeled 0-20, fraction II. The ammonium sulfate concentration was now raised to 30% and the resulting precipitate was recovered. This fraction was labeled 20-30, fraction III. In a similar fashion 30-40, fraction IV; 40-50, fraction V; 50-60, fraction VI; and 60-80, fraction VII were all prepared.

The various fractions were assayed and the results of a typical preparation are shown in Table 2-1. These data show that the fraction precipitating between 40 and 50% saturation contained 65% of the original activity. The fractions were assayed by adding 0.1 mL fraction, 0.1 mL luciferin (0.1 mg/mL), and 0.1 mL 0.01 M MgSO4, with 0.1 mL of ATP containing 1×10^{-8} g injected.

An attempt was made to achieve further purification by passing the 50-60 fraction through columns of various grades of Sephadex. Although in some cases some significant purification was attained, the recoveries were poor. Another reason for the attempt at further purification was to abolish the inherent light. However, this approach proved to be cumbersome and was not effective. Inherent light was reduced to a consistent low value simply by preparing a premix, as described later in this chapter, and allowing the inherent light to decay.

Luciferin. The luciferin used in these experiments was synthesized by Mr. George Svarnas, formerly of Fort Detrick, and by the McCollum-Pratt Institute at the John Hopkins University. The pure luciferin was stored until needed in several ampules under nitrogen in the deepfreeze. Each ampule contained 5 mg of pure luciferin. The effect of luciferin on light production is shown in Figure 2-3.

Adenosine triphosphate. Adenosine triphosphate was in the form of disodium trihydrate, crystalline, Sigma grade 99-100%. Five milligrams of ATP were diluted to 50 mL with pH 7.8 glycylglycine buffer; further dilution was carried out as needed.

Table 2-1. Purity of Luciferase Fractions.

% as Precipitate	Volume (ML)	Activity (mV)	Total Units	Recovery (%)	MG Protein/ML	Units/MG Protein	Relative Purity
1 A	50	1,450,000	72,500,000	100	26	55,800	1
0-20	10	265,000	2,650,000	3.7	27	9,810	1
20-30	10	225,000	2,250,000	3.1	19	11,800	1
30-40	10	624,000	6,240,000	8.6	10	62,400	1.1
40-50	10	4,740,000	47,400,000	65.4	25	190,000	3.4
50-60	10	170,000	1,700,000	2.3	13	13,100	1

AInitial extract.

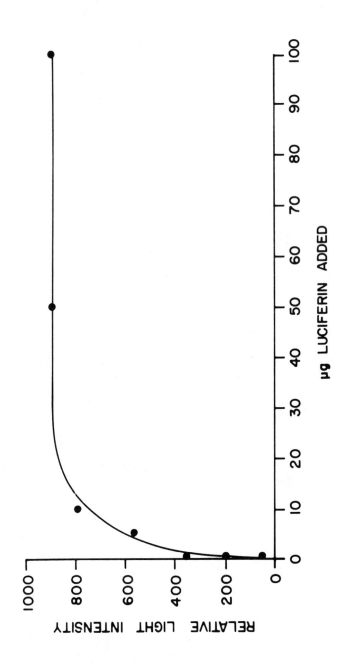

Figure 2-3. Effect of luciferin on light production.

It was important to be able to use the reagents, partially purified luciferase and luciferin, at conditions that would produce the minimum inherent light. The procedure has been standardized in the following manner. Luciferin was added to a partially purified preparation of luciferase. When this was done, the inherent light immediately rose to a high level and then decreased rapidly. After 2-36 hrs., the amount of inherent light decreased to a very low level. This decay in inherent light resulted in no loss of sensitivity with respect to ATP. Figure 2-4 shows the effect of time on the decay of inherent light.

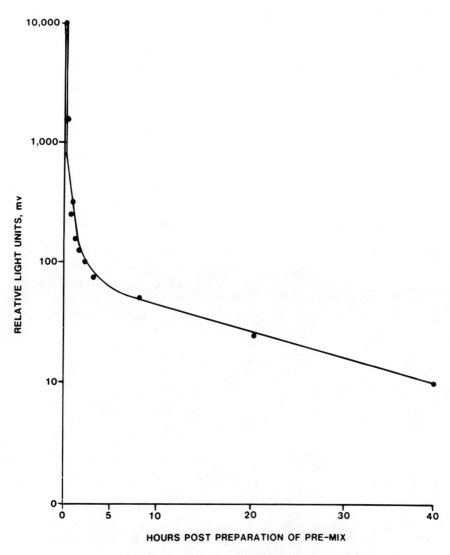

Figure 2-4. Effect of time on the decay of inherent light.

After the decay of the inherent light, ATP or samples containing ATP could be measured in the following manner. The premix was prepared by adding equal volumes of the luciferin, luciferase, and Mg^{+2}. This mixture was never frozen. When it appeared that sensitivity was being lost, the current premix was discarded and a new one prepared. Generally speaking, a premix lasted about a week.

Premix, 0.1 ml, was placed in a test tube that could be accommodated in a photometer. The instrument was allowed to record the inherent light for 1 min. At that point ATP or any unknown sample was added, usually by injection through a rubber stopper, and the initial flash height measured. These data are shown in Figure 2-5. Point A represents the point of injection of ATP or unknown; point B is a peak that always appears and then rapidly gives away to point C. Point C is the peak height that is measured. Point D represents the decay of flash height.

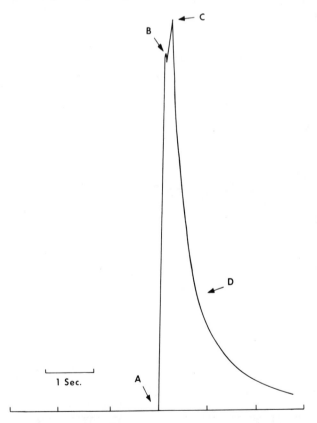

Figure 2-5. Initial flash height. At A the ATP or unknown is injected. The flash of light is characterized by an initial peak (B) and then a second peak (C). At all times point C has been used for measurement.

C. Detection of Microorganisms

A suspension of microorganisms was treated with perchloric acid to release the ATP. Any precipitable material was centrifuged and the supernatant fluid saved. The precipitate was washed once with buffer and added back to the original supernatant fluid. The combined supernatants were neutralized and any precipitate was discarded. The supernatant was then assayed for ATP in the same manner as described for a known solution containing ATP. Table 2-2 shows the ATP content of some microorganisms. Figure 2-6 demonstrates the linear response produced by the ATP in Serratia marcescens.

No attempt has been made to devise an automated instrument for the detection of microorganisms in the atmosphere because of difficulties in devising a system in which the luciferase does not become denatured.

III. CHEMILUMINESCENCE

Chemiluminescence can be defined as the light, usually visible, that is emitted as part of the energy released during certain exothermic reactions. There are a variety of substances that exhibit chemiluminescence and the subject has been well reviewed [15]. This review deals only with the use of luminol (Figure 2-7) because it is the only chemiluminescent substance that has been used for detecting small numbers of microorganisms. The luminescence of luminol (actually, the emitting species is the 3-aminophthalate ion), Figure 2-8, is dependent not only on the hemin moiety but also on a highly alkaline medium. The system works well for the detection of microorganisms. When bacteria are placed in a highly alkaline media (pH 13.0), they lyse and the hemin moiety is released into the reaction mixture.

Table 2-2. The ATP Content of Some Microorganisms.

Organisms	Mean ATP Content in uG ATP per Viable Cell
E. coli	2.37×10^{-16}
P. tularensis (LVS Strain)	9.70×10^{-15}
S. marcescens (PP Strain)	3.10×10^{-17}
S. marcescens (NSM-1 Strain)	3.32×10^{-16}
B. globigii spores	6.13×10^{-18}

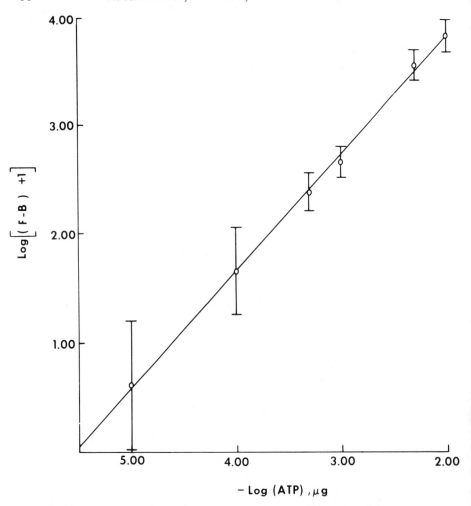

Figure 2-6. The linear response produced by ATP in Serratia marcescens. Flash height (F); Blank (B).

Figure 2-9 shows the linear response of luminescence produced by a fixed amount of luminol to varying small amounts of hemoglobin. Any changes in the nature of the hemin moiety produce significant alterations in sensitivity.

Table 2-3 shows a comparison between various iron-containing proteins. It is very apparent from these data that maximum sensitivity is obtained only when iron is present as hemin. Table 2-4 presents data showing numbers of microorganisms that can be detected. Figure 2-10 provides data demonstrating that the sensitivity in terms of the amount of detectable iron is very nearly the same in all substances that contain hemin as the prosthetic group.

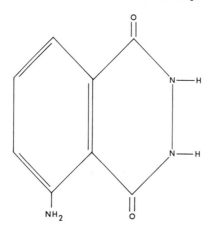

Figure 2-7. Luminol, 5-amino-2, 3-dihydro-1, 4-phthalazinedione.

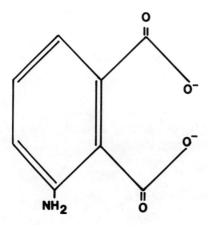

Figure 2-8. 3-Aminophthalate ion.

The assay system adopted by Vasileff [16] and Neufeld et al [17] was extremely sensitive. The unknown sample or hemin, alkaline luminol, and ethylenediaminetetraacetate were placed in a small test tube. The tube was placed in a dark chamber with an aperture facing a phototube. Hydrogen peroxide was injected into the tube through a rubber diaphragm and the light as measured by the phototube was recorded.

Assay systems for the detection of microorganisms can be made very sensitive, as indicated by the presented data, but caution must be exercised because impurities can give false signals. The use of a chemiluminescent

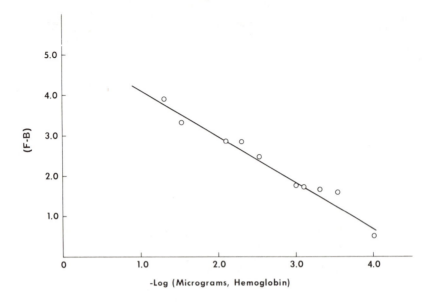

Figure 2-9. The linear response in terms of the amount of luminescence produced by a fixed amount of luminol to varying small amounts of hemoglobin.

system has the single advantage of not being dependent on the stability of an
enzyme preparation (luciferase) and is highly sensitive to a single catalyst
(hemin) that is almost universally distributed.

IV. ADAPTATION OF CHEMILUMINESCENCE TO DETECTING

MICROORGANISM AEROSOLS

Detection of microorganisms using a luminol system (5-amino-2, 3-dihy-
dro-1, 4-phthalazinedione) has been applied to clinical [18] and
environmental [19] samples. The amount of light emitted was found proportional to the number of bacteria present. Sensitivities of 10^3 to 10^6
organisms per milliliter were reported [20].

The possibility of applying this technique to the detection of aerosols of
microorganisms exists. Preliminary work has demonstrated that much more
needs to be accomplished to distinguish between background material and
live aerosols of microorganisms.

Table 2-3. Sensitivity Based on Iron Source and Content

Compound	Grams Detected	Molecular Weight	Moles Detected	Atoms of Iron Per Mole	Grams of Iron Detected
Ferritin	2×10^{-6}	750,000	2.7×10^{-12}	2500	3.9×10^{-7}
Cytochrome C	2.5×10^{-8}	13,000	1.9×10^{-12}	1	1.1×10^{-10}
Myoglobin	1.0×10^{-10}	17,000	5.9×10^{-15}	1	3.3×10^{-13}
Hemoglobin	1.0×10^{-10}	64,000	1.6×10^{-15}	4	3.6×10^{-13}
Hematin	6.0×10^{-12}	652	9.2×10^{-15}	1	5.2×10^{-13}
Catalase	1.0×10^{-10}	250,000	4.0×10^{-16}	4	9.0×10^{-14}

Table 2-4. Hematin Content of Various Microorganisms

Organisms	G Hematin/Viable Cell
S. marcescens	4.26×10^{-17}
P. tularensis	1.78×10^{-17}
P. pestis	4.82×10^{-17}
B. anthracis spores	2.55×10^{-15}
E. coli	1.74×10^{-16}
S. aureus	1.18×10^{-17}
M. lysodeikticus	3.61×10^{-17}
B. subtilis var. Niger spores	8.23×10^{-18}
B. cereus spores	4.10×10^{-16}

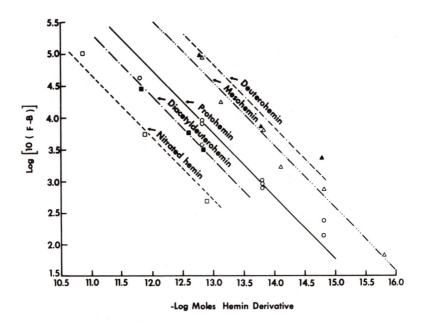

Figure 2-10. Significant alterations in sensitivity produced by changes in the nature of the hemin moiety.

V. REFERENCES

1. Lehninger, A. L., "Biochemistry"; 2nd ed.; Worth Publishing Co.: New York, 1975; p. 509.
2. McElroy, W. D.; Seliger, H. H., "Bioluminescence in Progress"; Princeton University Press: Princeton, NJ, 1966; p. 432.
3. Seliger, H. H.; McElroy, W. D. Biochem. Biophys. Res. Commun., 1960, 1, 21.
4. Neufeld, H. A.; Towner, R. V.; Pace, J. Experientia, 1975, 31, 391.
5. Prydz, H.; Froham, L. D. Acta Chem. Scand., 1969, 18, 559.
6. Beutler, E.; Buluda, M. C. Blood, 1964, 23, 668.
7. Strehler, B. L.; Totter, J. R. Fed. Proc., 1952, 11, 295.
8. Holmsen, H.; Holmsen, I.; Bernhardsen, A. Anal Biochem., 1966, 11, 456.
9. Cole, H. A.; Wimpenney, J. W. T.; Hughes, D. E. Biochem. Biophys. Acta, 1967, 143, 445.
10. Holm-Hansen, O.; Booth, Cr. R. Limol. Oceanogr., 1966, 11, 510.
11. Van Dyke, K.; Stitzel, R.; McClellan, T.; Szustrienicz, C. Clin. Chem., 1960, 15, 3.
12. Stanley, P. E.; Williams, S. G. Anal Biochem., 1969, 29, 381.
13. St. John, J. R. Anal Biochem., 1970, 37, 409.
14. Hammerstedt, R. V. Anal Biochem., 1973, 52, 449.
15. Cormier, M. J.; Hercules, P. M.; Lee, J. "Chemiluminescence and Bioluminescence"; Plenum Press: New York, 1913.
16. Vasileff, T. P.; Svarnas, E.; Neufeld, H. A.; Spero, L. Experientia, 1974, 30, 20.
17. Neufeld, H. A.; Conklin, J.; Towner, R. D. Anal Biochem., 1965, 12, 303.
18. Ewetz, R.; Strangert, K. Acta Pathol. Microbiol. Scand., Sect. B, 1974, 82, 375.
19. Oleniacz, W. S.; Pisano, M. A.; Rosenfeld, M. H.; Elgart, R. L. Environ. Sci. Technol., 1968, 2, 1030.
20. Miller, C. A.; Vogelhut, P. O. Appl. Environ. Microbiol., 1978, 35, 813.

3. RAPID MICROBIOLOGICAL ANALYSIS

BY FLOW CYTOMETRY

W. Keith Hadley, Frederic Waldman, and Mack Fulwyler

I. INTRODUCTION

Flow cytometry is a highly effective means for rapid analysis of individual cells at rates of up to 1000 cells per second. The fundamental analytical systems are fluorescence and light scattering [1-3]. By staining cells with specific fluorochromes or fluorescent conjugates, it is possible to measure a wide variety of cell constituents, such as proteins, carbohydrates, DNA, RNA, enzymes, and antigens [3a]. Improvements in staining and instrumentation have allowed simultaneous measurement of several parameters [4-6], and the capacity to sort out droplets containing cells with specific known characteristics [7-10]. Conventional biochemical analyses of cell samples measure the average character of the entire cell population. Flow cytometry provides measurement of individual cells. The variation in cell composition can be determined and subsets of a cell population can be detected. The study of mammalian cells using flow cytometers has been particularly fruitful, resulting in information on cell kinetics [11], tumor cell characteristics [12-14], separation and characterization of chromosomes [15-17], separation of lymphocytes into different functional subsets with different surface marker phenotypes [18-20], phagocytosis of bacteria [21], and localization of receptors for bacterial structures on cells [22]. Cram described [23] the potential of flow cytometry for analysis of viruses and bacteria at the Second International Symposium on Rapid Methods and Automation in Microbiology in 1976. He pointed out that because of their small size and the low number of DNA molecules present to be stained—typically three orders of magnitude less than a mammalian cell—techniques for quantitating properties of single bacteria were difficult and required instruments and methodology of high sensitivity and sophistication. Relatively few investigators used flow cytometry for the study of prokaryotic microorganisms prior to 1980. This is unfortunate because the intrinsic advantage of flow cytometry, the rapid (1000 cells per second) analysis of individual cells, has not been used adequately.

A marked increase in the sensitivity and specificity of the tools of flow cytometry has occurred in the past few years. Consequently, microbial analysis is feasible. Recent developments that enhance the usefulness of flow cytometry for microbiology include dual-beam flow cytometers, with high-energy lasers for excitation of fluorochromes; new fluorochromes of high intrinsic fluorescence output; and the capacity to demonstrate specific ligands by fluorescent-dye-tagged reagents, such as fluorogenic substrates for enzymes, or fluorochrome-tagged antibodies, lectins, or complementary RNA or DNA.

Flow cytometry offers the investigator a powerful new analytical instrument that is quite adaptable to the techniques of molecular biology. Flow cytometry is potentially applicable to analyses in clinical, environmental, and industrial microbiology. Present analytical systems depend upon a microbial growth phase and are therefore slow. Direct analysis of each cell could make microbial analysis more responsive to the need for rapid diagnosis and therapy in the clinical setting. Environmental microbiology and industrial process control done by flow cytometry can be rapid processes and therefore are more likely to be used. The flow cytometry of microorganisms and viruses has entered a highly active period of research and application.

This chapter describes the characteristics of flow cytometers, sorters, and associated analytical systems, as well as their application to problems of microbiology.

II. GENERAL DESCRIPTION OF FLOW ANALYTICAL SYSTEMS

The advantage of flow cytometry lies in its ability to make rapid, quantitative measurements of multiple parameters of each cell within a large number of cells; this makes it possible to define the properties of the overall population and of component subpopulations. This technology allows correlated multiparameter analysis of each cell, combining information concerning intrinsic cell properties, such as volume or light-scattering characteristics, with user-defined properties dependent on staining with specific fluorescent probes. The reader who would like to begin to use flow cytometry is directed to the description of "flow sorters" by Herzenberg and Sweet [10], and the multiauthored book edited by Melamed et al [24].

A. Sample Preparation and Fixation

Cell aggregates can be disrupted by enzymatic treatment, eg, by proteases or DNAase, or by mechanical disruption, eg, gentle pipetting through small-bore needles or more vigorous sonication [25]. Although frequently used, fixation is not required for flow-cytometric analysis, and in some cases it may interefere with fluorescent labeling. The advantages of fixation are greater stability of the sample (days to months), increased permeability of the cell membrane to large molecules (antibodies and certain fluorochromes), and the killing of pathogenic organisms. Fixation with crosslinking reagents (paraformaldehyde, glutaraldehyde) may require induction of greater permeability with detergents for adequate staining, whereas lipid-soluble fixatives (ethanol, methanol, acetone) may cause increased aggregation of cells [26,27].

B. Fluorescent Probes

The fluorochromes more frequently used in flow-cytometric analysis are listed in Table 3-1. At saturating concentrations, dyes will specifically bind to nucleic acids, proteins, or other molecules stoichiometrically, allowing quantitative measurement based on fluorescence intensity [28]. Fluorescent probes also have been utilized to measure membrane potential [29], intracellular pH [26], enzymatic activity [26], and membrane integrity or viability

Table 3-1. Fluorescent Stains Used in Flow Cytometry

Fluorochrome	Uses	References
Acridine orange	DNA and RNA intercalation; double strands, green; electrostatic single strands, red	27, 37
Chromomycin A3	Guanine-cytosine-rich DNA	38
DiOC^1(3)-cyanine dye	Membrane potential	29
Ethidium bromide	RNA/DNA, DNA after RNAase; cell viability	37, 38
Fluorescein isothiocyanate (FITC)	Total protein; conjugates with antibody, lectins, hormones	35, 39, 40
Fluorogenic substrates	Detection of enzymes	26
Fluorescein diacetate (FDA)	Cell esterase, cell membrane permeability and viability	29, 30, 31
Hoechst 33258	Adenine-thymine-rich DNA, cell membrane permeability and viability	37, 41, 42
Hoechst 33254	Cell membrane permeability and viability	32
Mithramycin	DNA	38
Phycobiliproteins (phycocyanines, phycoerythrin)	High-intensity fluorescence; conjugates with ligands, cofluorophore with FITC	43, 44
Propidium iodide	RNA/DNA, DNA after RNAase, cell membrane permeability and viability	37, 38
Texas red	High-intensity fluorescence; cofluorophore with FITC, conjugates	45

[29-31]. In addition, biologically active molecules, such as hormones, toxins, colonizing or adherence factors, and antibiotics, can be tagged fluorescently, or a fluorescent conjugate can be used to locate the ligand or receptor. Fluorescein-lectin conjugates, which may bind to cell surfaces, have been used to identify human monocytes [32] and as a means to separate mutant yeast cells useful for studying mannoprotein synthesis using the fluorescence-activated cell sorter [33]. Specific immunofluorescent probes may be selected from polyclonal, affinity-column-purified antibody of high avidity, or highly specific monoclonal antibody. These antibodies can be conjugated to a variety of fluorescent dyes [34]. Fluorescein is used most frequently, although many other fluorescent markers are available. Fluorochromes with distinguishable emission spectra can be selected to measure simultaneous binding of different antibodies. A fluorochrome-conjugated anti-immunoglobulin-immunoglobulin sandwich or a hapten sandwich, such as biotinylated antiimmunoglobulin, which binds to fluorochrome-conjugated avidin, serves to produce intensified fluorescence as compared to the fluorescence caused by the direct fluorochrome-conjugated immunoglobulin reaction with antigen [35].

C. Laminar Flow

The basic technique of flow cytometry requires that cells in aqueous suspension be focused into a narrow stream and then passed through a sensing region at high speed. The sample must be in a single-cell suspension to permit analysis of single cells, and to avoid clogging of the sample orifice by cell aggregates. The sample is injected into a larger stream of flowing "sheath" fluid, and the resultant process of hydrodynamic focusing insures smooth laminar flow (Figure 3-1), narrowing and centering the sample stream for precise optical measurement of individual cells. Approximately 0.01 mL of sample fluid surrounded by 1.5 mL of sheath fluid pass through the orifice per minute at a velocity of 1-10 m/s. This allows high-speed analysis and sorting of cell populations.

D. The Photodetection System

As the liquid jet emerges from the orifice it passes through a light beam that has been focused onto the sample stream (Figure 3-2). Each cell "scatters" a small amount of laser light as it passes through the incident beam, and this light can be focused onto a photodetector, generating an electrical signal that is proportional to the intensity of scattered light. Light that is collected at small angles from the incident beam, in the range of 1-10°, is a good measure of the cell size. The amount of scattered light collected at right angles to both the incident laser beam and the liquid jet is dependent on intracellular structure. The presence of granules within polymorphonuclear leukocytes, for example, allows their discrimination from lymphocytes on the basis of right-angle scatter.

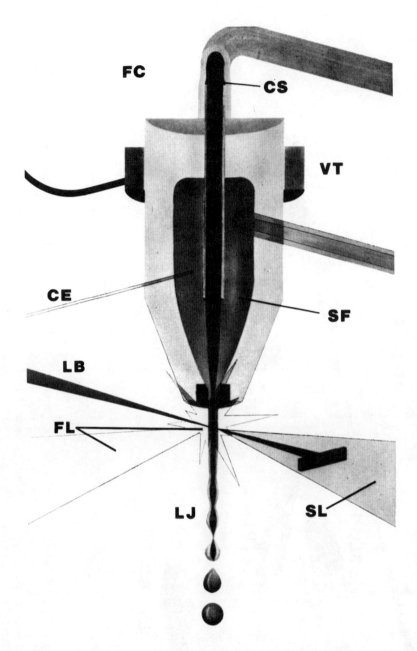

Figure 3-1. Laminar flow cell. FC, flow cell; CS, cell suspension; VT, vibration transducer; CE, cell electronic feedback control; SF, sheath fluid; LB, laser beam; FL, fluorescent light; SL, scattered light; LJ, liquid jet. (Courtesy of Becton Dickinson FACS Systems.)

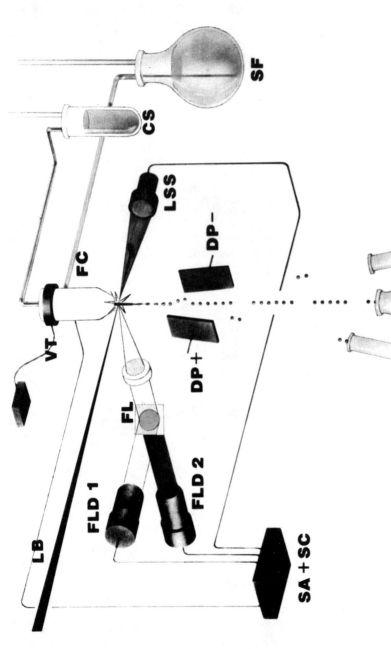

Figure 3-2. Cell sorter. FC, flow cell; VT, vibration transducer; CS, cell suspension; SF, sheath fluid; LB, laser beam; FL, fluorescent light; FLD 1, fluorescence detector 1; FLD 2, fluorescence detector 2; LSS, scattered light sensor; DP(+), positive deflection plate; DP(−), negative deflection plate; SA+SC, sorter analyzer + sorter control. (Courtesy of Becton Dickinson FACS Systems.)

Although most flow-cytometric systems use low-angle light scattering to measure cell size, some instruments use the Coulter principle. This method of volume measurement is based on the fact that a cell passing through the orifice in a stream of conducting fluid causes a drop in DC conductance that is dependent on cell size. The accuracy of measurement is very high and particles as small as platelets are conveniently measured. Measurement of bacteria is possible but difficult.

Fluorescence detection is a widely used tool in flow-cytometric analysis. Because excited fluorescent material emits light in all directions, a photo-detector at right angles to the exciting light beam can measure cell-associated fluorescence. Appropriate filters can screen out all wavelengths but those of the emitted fluorescence. If more than one fluorescent probe is being used, the emitted fluorescence can be split and filters used to pass selectively to separate photodetectors light associated with each probe.

Although commercial instruments using mercury-arc illumination exist, a laser light source generally is used for light scattering and fluorescence excitation of microbiological samples because of greatly enhanced sensitivity. This occurs because the laser beam can be focused to a small cross section, and the beam is of very high intensity. Each type of laser emits light at several discrete wavelengths, which allows for excitation of a variety of fluorescent dyes. Some instruments are equipped with two laser light sources, allowing measurement of combinations of fluorescent dyes with distinguishable excitation spectra (Figure 3-3). Most useful laser light sources at this time are expensive and require continuous water cooling and a great amount of electrical power. Instruments that use mercury-arc illumination provide less sensitivity of measurement than those with laser illumination; however, their lower cost, greater flexibility, and simplicity of use make them valuable tools for both research and clinical practice.

E. Cell Sorting

Although most uses of flow cytometry require only analysis of the stained population, sorting of preselected subpopulations also can be achieved with some intruments. After the liquid jet has traversed the sensing region, the action of a vibrating oscillator mounted to the nozzle breaks up the jet into numerous small droplets, usually at a frequency of 40,000 per second. If a cell selected for sorting falls within preset size and/or fluorescence criteria, then the drop likely to contain that cell, along with the preceeding and following drops, are charged electrically as they are formed. As these drops pass between two high-voltage plates, they are deflected from the remaining stream of uncharged drops. Drops can be given a positive charge or a negative charge, causing deflection to either side. The deflected charged drops and undeflected neutral drops are then collected into separate containers. Because 1000-5000 cells can be analyzed per second, and drops are formed at a much higher rate, a drop rarely contains more than one cell. If this type of "coincidence" of cells does occur, the "coincidence" can be detected and those drops can be excluded electronically from the sorted population. The entire sorting system can be kept sterile and can be modified easily to prevent aerosol formation when biohazardous materials are being analyzed.

F. Data Analysis

Most flow-cytometric systems allow simultaneous measurement of cell volume and two or more fluorescence signals. When sorting, "selection windows" can be set electronically to select only those cells with established size or fluorescence characteristics. Further analysis of the data usually requires computer-assisted processing. Data can be stored as multiparameter-correlated sets (ie, all measurements for each cell are "listed" together) and plotted at a later time. Usually the data are first "gated" for specific criteria of one parameter, for example, only cells falling within a certain size range are selected for further analysis; then two fluorescence parameters may be obtained and plotted one against the other. Plotting of data on log-log scales is frequently helpful when a broad fluorescence intensity range is

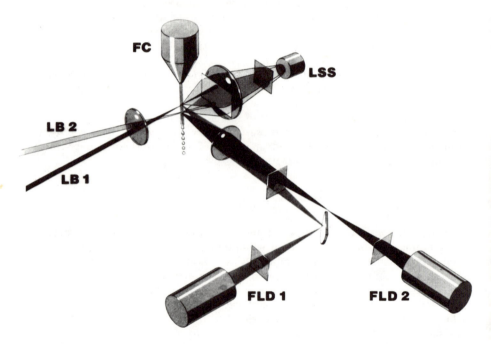

Figure 3-3. Photodetection system, dual laser beam flow cytometer. FC, flow cell; LB 1, laser beam 1 (focused by mirrors and lens); LB 2, laser beam 2; LSS, scattered light sensor (aperture, lens, filter, photodetector); FLD 1, fluorescent light detector 1 (obscuration, lens, filter, beam splitter, filter, photodetector); FLD 2, fluorescent light detector 2. (Courtesy of Becton Dickinson FACS Systems.)

present, or when the cell population being studied has a log-normal distribution. Statistical analysis of flow data is aided by the large number of cells normally measured. Homogeneous cell populations should exhibit a small coefficient of variation, certainly less than 5%, for small differences in fluorescence intensity between cell populations to be detected. Data from multiparameter analyses usually indicate frequency by contour lines (Figure 3-4) or by simulated depth.

Manufacturers of flow-cytometry instruments are listed in Table 3-2.

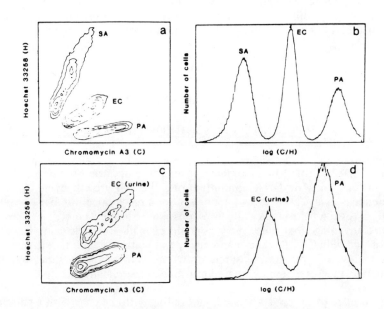

Figure 3-4. (a) Bivariate contour plot of the fluorescence signals from a mixture of cultured Staphyloccus aureus (SA), Escherichia coli (EC), and Pseudomonas aeruginosa (PA) after double DNA staining with Hoechst 33258 (H) and chromomycin A3 (C). The peak height (number of cells) is indicated by the contours. (b) Distribution of the ratio of the two fluorescent signals, C/H, over the cell population of (a). (c) Bivariate contour plot of the fluorescence signals from E.coli from an infected urine sample after double DNA staining with Hoechst 33258 and chromomycin A3. Cultured P. aeruginosa was added as an internal standard. (d) Distribution of C/H over the cell population of (c). (Reproduced with permission from ref. [59]. Copyright 1983 by the AAAS.)

Table 3-2. Manufacturers of Flow-Cytometry Instruments

Becton Dickinson FACS Systems Division, 490-B Lakeside Drive, Sunnyvale, California, 94086. Laser-based instruments of the FACS series and mercury-arc-based instruments of the Research Analyzer series are available.

Coulter Electronics, Inc., 590 West 20th Street, Hialeah, Florida, 33010. Coulter Electronics offers the EPICS series of laser-based instruments.

Kratel GmbH & Co KG, Boeblinger Strasse 23, D-7250 Leonberger-Stuttgart, West Germany. This firm offers laser-based flow cytometers.

Ernst Leitz Wetzlar GmbH, D-6330 Wetzlar, West Germany, offers a mercury-arc flow-cytometry accessory to their fluorescence microscopes.

Ortho Diagnostic Systems, Raritan, New Jersey, 08869. Two lines of laser-based instruments are available, the Cytofluorograph line and the Spectrum III line. A mercury-arc-based instrument is also available.

III. ANALYSIS OF CELL MASS AND NUCLEIC ACID

DURING THE CELL GROWTH CYCLE

The first uses of flow cytometry in microbiology were to study cell mass and nucleic acids or DNA content during the cell growth cycle. Standard biochemical analyses of these parameters required a large number of cells in as close to a synchronous culture as possible. Flow cytometry offered the distinct advantage that these analyses could be made on single cells. The qualities of individual cells that made up a population could be studied. The earliest flow cytometers measured only one character at a time, but later cytometers measured and correlated multiple measurements per cell.

Paau et al [46], used scattered light falling within a cone with a half-angle of 20°, centered at 90° from the forward direction as a measure of cell mass. Ethanol-fixed cells were stained with ethidium bromide, which intercalates between the double-stranded nucleic acid. Under the conditions of their study, they found that deoxyribonuclease removed the fluorescent staining from the bacteria they analyzed. They found that Escherichia coli with a relatively short generation time has eight-fold greater DNA and three-fold greater cell mass than stationary phase cells. Rhizobium japonicum, with a ten-fold greater generation time, had two-fold or less DNA per cell in log phase as compared to stationary phase.

The same investigators applied their flow-cytometry method to a problem on legume root nodule Rhizobium bacteroids, which do not grow on nutrient medium [47]. The bacteroid nonviability had been ascribed to a decrease in the nucleic acid content in the symbiotically growing Rhizobium. Only small numbers of bacteroids could be extracted from root nodules. Paau and co-workers found that, in fact, the small number of bacteroids available for analysis had a greater nucleic acid content per cell than Rhizobium growing in vitro.

Bailey and co-workers [48] studied Bacillus subtilis batch cultures with flow cytometry. They used ethanol-fixed cells stained with fluorescein isothiocyanate (FITC) and propidium iodide (PI). The fluorescence of FITC was an effective measure of cell protein and PI fluorescence measured double-stranded nucleic acid. Protein content varied little from cell to cell in the spore suspension inoculum but increased and had greater variance during log growth when the spores germinated. A bimodal result was present in late cultures when spores and vegetative cells were present. Cells in the log phase had much variation in nucleic acid content per cell.

Hutter and Eipel [49] studied the yeast Saccharomyces cerevisiae with flow cytometry using FITC stain for protein and propidium iodide stain for DNA. They excited fluorochromes at 488-nm wavelength and separately measured the green fluorescence of FITC and the red of PI. They found that the FITC stain contributed a red fluorescence, which produced an increasing red signal as protein increased. This could be adjusted by changing the FITC stain concentration and by computer adjustment of data based on an FITC control. This result was more noticeable in yeasts that had a comparatively small amount of DNA per cell than in mammalian cells.

Studies of yeast nucleic acid and protein also have been used to determine the ploidy level of Candida albicans [50], changes in yeast cells metabolizing glucose and ethanol [51], and in cell cycle analyses of batch cultivation of yeast [52].

In a series of reports on flow cytometry of bacterial growth, Steen and co-workers [53-56] studied bacterial cell cycle kinetics. They used ethanol-fixed cells, with ethidium bromide and mithramycin staining of nucleic acid, and fluorescein isothiocyanate (FITC) staining or light scattering as a measure of cell protein or "cell mass." These studies were done with a fluorescence microscope-based flow cytometer constructed by Steen and Lindmo [57]. The exciting light in this instrument was obtained with a 100-W mercury-arc lamp or with a 5-W argon laser.

Using this sytem and dual-parameter analysis, they showed that there was a linear relationship between the intensity of scattered light and the fluorescence intensity of individual bacteria stained by FITC. The coefficient of variation in these measurements was 12-14%. In an initial study to calibrate the fluorescence from cellular DNA they studied E. coli K12, strain E177. When grown at $42^{\circ}C$, this strain is defective in initiating new rounds of DNA replication but can divide. After prolonged growth, these bacteria had fluorescence from the DNA of one full chromosome per cell.

Flow-cytometer histograms showed two peaks, with the lower peak representing cells having the lowest DNA content, ie, cells that had replicated one chromosome to two chromosomes, and the higher value histogram peak represented four full chromosomes. Bacteria in rapid exponential growth just prior to the stationary phase, when the number of bacteria no longer increased, contained DNA equivalent to two full chromosomes. Only upon continued incubation overnight at 37°C did the expected single chromosome peak occur in a sizeable proportion of the bacteria. When experiments were repeated with exponentially growing cultures, it was not possible to obtain exactly the same DNA/light-scattering histograms. When batch cultures or chemostat cultures were employed, there was significant variation in the histograms.

These flow-cytometry experiments show the complexity and variability of cell populations at various phases of the growth cycle. These complexities could not be appreciated in the total biochemical analyses of masses of cells.

IV. ANALYSIS OF ANITIMICROBIAL ACTION ON BACTERIA

The mechanism of antimicrobial action on bacterial cell growth and DNA replication has been studied using flow cytometry. The potential for studies in this field is great. For example, antibiotics conjugated with fluorochromes could be used to study the presence of antibiotic receptors and cell binding. Also, immunofluorescent methods may be used in similar fashion to detect cell receptors. Fluorogenic substrates that bind to cells may detect antibiotic metabolizing enzymes present on individual cells.

Steen et al [55] and Boye et al [56] using the flow-cytometry methods they developed to analyze the DNA complement of E. coli during the cell growth cycle, have studied the action of antimicrobials that inhibit protein synthesis. Mithramycin and mithramycin-ethidium bromide staining of DNA was used in these respective studies. At 100 μg/mL doxycycline completely stopped cell division and further DNA synthesis. The block was permanent. No change occurred when the antimicrobial was removed. The histograms of E. coli treated with 100 μg/mL of chloramphenicol showed two prominent peaks of cells with two and four completely replicated chromosomes, respectively. Streptomycin with 100 μg/mL and erythromycin with larger doses produced similar results. These results indicated inhibition of initiation of DNA synthesis, but DNA synthesis that already had started proceeded to completion.

V. DOUBLE FLUORESCENT STAINING TO DETERMINE DNA

BASE PAIR RATIO (% GC) BACTERIAL CHARACTERIZATION

Van Dilla et al [59] have applied to bacteria a double fluorescent DNA staining method that originally was developed by Langlois et al [60] for analysis of mammalian chromosomes. Chromomycin A3(C) binds preferentially to guanine-cytosine (GC)-rich DNA and Hoechst 33258 (H), has an adenine-thymine (AT) preference. The stain combination provides information on the DNA base composition (proportion of GC and AT base pairs) as well as the total DNA content. Bacterial species vary widely in DNA base composition. Guanine-cytosine values range between 25 and 70% of the total base pairs in the DNA and have been used as one of the main tools in classification and taxonomy of bacteria. In these studies, bacterial cells were concentrated by centrifugation, resuspended in 70% ethanol, and held at $4^{o}C$. Small portions of fixed bacterial suspension were added to a staining solution containing Hoechst 33258, chromomycin A3, tris buffer (pH7.2), 150 mM NaCl, and 1.5 mM $MgCl_2$, usually in a concentration of 10^7 to 10^8 bacteria per milliliter. The two fluorescence emissions from the bacteria were measured with a dual-beam flow cytometer and correlated. A reference suspension of Pseudomonas aeruginosa was added to each stained sample so that fluorescence ratios could be determined relative to the standard bacteria and be independent of instrument gain settings.

Measurements of bacteria were recorded in a bivariate contour plot with fluorescence of chromomycin (C) on the horizontal axis and Hoechst 33258 (H) on the vertical axis (Figure 3-4). The radial coordinate was proportional to the cellular DNA content and reflected chromosome replication, incompletely separated cells, and possibly clumping. The angular displacement of the pattern from the vertical (H) axis is a function of the GC percentage. Different bacterial species present in a mixed sample can be distinguished and could be sorted if this were useful. Significant differences in the C/H ratios were found for different strains of E. coli (0.31-0.41). In subsequent studies [61] bacterial strains whose percentage of GC had been measured by DNA melting temperature (percentage GC measurements made by D. Brenner, A. Steigerwalt, and J. L. Johnson) were stained and measured by this system. Using logarithmic conversion of the data, a linear calibration curve can be obtained to relate the flow-cytometer-derived fluorescence of individual Gram-positive and Gram-negative bacteria to percentage GC for these bacteria (Table 3-3).

This simple fixation and staining system requires a dual laser beam flow cytometer and computer for analysis. The system offers a rapid and precise means of DNA analysis for taxonomic studies. Further evaluation of the relationship between conventional percentage GC and the C/H measurements is necessary. The extent to which the C/H measurement and total DNA can be used for rapid characterization of bacterial species awaits measurements of different strains. The influence of large plasmids or cells with numerous

Table 3-3. A Comparison of % GC Measured by Conventional Methods
with Fluorescence Ratios for Selected Bacteria

Bacterium	Percentage GC	Fluorescence ratio, C/H ($\pm$ 1 SD)
Pseudomonas aeruginosa		1.0
Propionibacterium acnes	59	0.62 ± 0.02
Klebsiella pneumoniae	55	0.46 ± 0.03
Escherichia coli	50	0.34 ± 0.02
Bacteroides fragilis	42	0.28 ± 0.02
Lactobacillus acidophilus	33	0.13 ± 0.01
Clostridium butyricum	28	0.072 ± 0.008

plasmid replicates is not known. Application of this flow cytometry system to characterization and enumeration of bacteria in infected urine specimens is described in Section XI of this chapter.

VI. IMMUNOFLUORESCENT STAINING OF BACTERIA

The immunofluorescent staining of bacteria followed by microscopic observation is a well-established method for rapidly detecting and identifying bacteria. This can be accomplished either with direct clinical or environmental samples, or with samples that have been allowed a brief growth period in order to increase the sensitivity of the test [35,39]. Between 10^4 and 10^5 fluorescently stained bacteria per millileter are necessary for direct microscopic detection on slide preparations. Flow cytometry has not been used frequently for detecting immunofluorescently stained bacterial cells. Improved specificity of reagents by use of affinity purified or monoclonal antibody and more intensely fluorescent fluorochromes for labeling of the antibody offer improved specificity and sensitivity.

Ingram et al [62] working with a Coulter EPICS IV Cell Sorter used the 488-nm line of an argon ion laser to excite the FITC of rabbit anti-Legionella pneumophila serotype 1 conjugate. They found that 95% of L. pneumophila serotype 1 cells were stained by the conjugate. Specific staining of L. pneumophila was obtained even when mixed with E. coli and Saccharomyces cerevisiae cells. The flow cytometer was able to detect bacteria at a concentration of 10^4 stained bacteria per milliliter, the lowest concentration tested.

Very low fluorescence intensity was obtained with distilled water and no organisms and with L. pneumophila and FITC-labeled normal rabbit globulin. A separate but broad peak of fluorescence was obtained when FITC-conjugated rabbit anti-L. pneumophila serotype 1 globulin was used to stain L. pneumophila serotype 1. This is compatible with the pleomorphism of L. pneumophila in vitro. The authors propose direct staining of environmental material and analysis by flow cytometry as a potentially useful presumptive test for L. pneumophila which avoids the tedious and time-consuming quantitation by microscopy.

Steen et al [55], using their fluorescent microscope-based flow cytometer, studied the binding of pooled human serum from Abe's lepromatous patients to suspensions of Mycobacterium leprae obtained from armadillo tissue. Binding of the human antibody to the M. leprae cells was detected by FITC conjugated to rabbit anti-human immunoglobulin antibody. The antibody binding appeared to be largely independent of cell size as measured by light scattering. The authors propose that the extent of binding of a standard antibody is a quantitative measure of antigen in a bacterial vaccine.

Phillips and Martin [63] studied immunofluorescent staining of Bacillus anthracis spores and vegetative cells. They used hyperimmune rabbit antisera to B. anthracis spores purified to IgG fractions and conjugated with FITC. Measurements were made with the Ortho Cytofluorograph 30L, operated at 30 mW of 488-nm illumination to excite the fluorescence. The fluorescence signals had components from unreacted conjugate, from complexes of conjugate with extracellular debris, and from soluble antigen, as well as from the stained cells themselves. In order to measure the fluorescence of stained bacteria, they proposed routine adjustment of the histograms by subtracting the fluorescence of the supernatant after centrifugation. They suggested further study of the complexity of the fluorescence signals by sorting the cells for analysis.

Waldman et al [64] have studied immunofluorescent staining of the obligate intracellular bacterium Chlamydia trachomatis using FITC-conjugated monoclonal antibody to Chlamydia surface antigen. Infected L cells were fixed with paraformaldehyde and permeability was increased with saponin. Quantitative titration of infected cells in vitro was accomplished rapidly by this technique. Use for diagnosis of human infection is being studied. Chlamydia directly conjugated to FITC may be used to study the mechanism of Chlamydia attachment to host cells.

VII. VIRAL STUDIES

Analysis of free viral particles and the measurement of viral attachment to sensitive cells is possible using flow-cytometric techniques. Although the light scattered by particles with a greatest dimension smaller than the wavelength of light used is not directly proportional to the particle size or shape, individual bacteriophage T2 and reovirus particles can still be detected by scattered light measurements [65]. Flow cytometry also has been used to measure the adsorption of fluorescently labeled viruses to eukaryotic cells [66]. Viral binding was both specific and saturable and showed a wide variation in the number of virions bound per cell.

Flow analysis has been used more widely to study virally infected cells. Changes in host cellular DNA content following infection by SV40 or adenovirus suggested effects on cell cycle kinetics [67,68]. Viral replication of herpes I in lytically or persistantly infected cell culture was detectable by DNA measurement [69]. Immunostaining for herpes II viral antigens has been used in cell-sorting experiments to show a correlation between cellular atypia and herpes markers in exfoliated cervical cells [70]. A combination of light scatter and light absorption of immunoperoxidase stained herpes-infected cells reliably detected infection [71]. Fluorescent monoclonal antibody staining of clinical specimens and tissue culture systems to diagnose viral infection is becoming more generally used, and applications of flow techniques are promising.

VIII. DETECTION AND ANALYSIS OF PROTOZOA

Flow cytometry has been used in particular for the study of protozoa that parasitize erythrocytes. More traditional methods, eg, microscopic examination of parasitized erythrocytes, which include studies of erythrocyte maturity and parasite differentiation in the erythrocyte, are laborious. Asynchronous and multiple infections make separation of parasites for antigenic and biochemical analysis difficult. In contrast, dual-parameter flow-cytometer analysis, using fluorescent dyes and scattered light combined with cell sorting, has been successful in separating erythrocytes containing parasites at various stages of maturity.

Jacobberger et al [72] studied the blood of mice infected with Plasmodium vinckei. The cyanine dye DiOC[1] [3] which measures membrane potential, distinguished subpopulations of normal and infected host cells. A comparison of light scattering versus fluorescence intensities is used to discrimate erythrocytes, platelets, granulocytes, and mononuclear cells. When combined with Hoechst 33342 staining, infected and uninfected erythrocytes, as well as mature and immature erythrocytes, could be discriminated. Howard et al [73], studying Plasmodium berghei-infected mouse blood and Plasmodium falciparum parasites from long-term vitro culture, used the DNA stain Hoechst 33258 for separation and sorting of infected and uninfected erythrocytes. The infected erythrocytes, which exhibit fluorescence in nearly direct proportion to the number of parasite nuclei present, are sorted easily because mature erythrocytes lack DNA staining. Hoechst 33258 has the additional advantage that with properly selected staining conditions in vitro, the infecting capacity of the P. berghie can be retained and studied. Whaun et al [74], studying continuous erythrocyte culture of P. falciparum, were able to discriminate ring forms, trophozoites, and schizonts with acridine orange stain. They were also able to isolate the various stages of infected cells with a cell sorter.

Howard and Rodwell [75] also used Hoechst 33258 staining of DNA for analysis and sorting of normal from infected red cells from Babesia-infected mouse and calf blood. They sorted on the basis of cell fluorescence and on its intensity of light scattered at low angles (related to cell size). Preliminary column treatment was necessary to remove leukocytes, a major source of DNA-stained material in whole blood. The vital staining quality of Hoechst 33258 allowed subsequent study of the infectivity of sorted Babesia from infected erythrocytes.

Flow cytometry has been used to detect the free-living protozoa, Naegleria and Acanthamoeba, which may infect humans, producing acute and chronic meningoencephalitis. Examination of water and soil samples for these protozoa is described in Section X.

IX. SINGLE-CELL PLANT STUDIES

Flow cytometry also has been applied to the study of unicellular plants. Trask et al [76] studied forward and perpendicular scattered light intensities, chlorophyll fluorescence excitation spectra, and DNA staining by Hoechst 33342 and DAPI of phytoplankton algae. These combined optical measurements enabled the distinction of several algal species. The authors note that flow cytometry may prove to be an efficient means for measuring water quality by quantification of algal content. Bonaly and Mestre [77] have also used flow cytometry to study algae. They studied the cell cycle of the freshwater alga Euglena gracilis.

X. USE OF FLOW CYTOMETRY FOR ENVIRONMENTAL MICROBIOLOGY

Environmental microbiology tests are used to insure that standard public health safety measures are operating effectively. In some cases, it is necessary to detect an infectious agent that may be present in air, water, soil or food, and to find its source and mode of transmission. Flow cytometry has the potential for rapid detection and characterization of microorganisms in these applications. Immunofluorescent staining has been used to detect two significant groups of soil and water organisms that are capable of causing serious disease in humans: the Legionella and the free-living protozoans Naegleria and Acanthamoeba.

Direct immunofluorescent staining of <u>Legionella</u> and flow-cytometric analysis described by Ingram (see Section VI) should be tried on true environmental and clinical specimens. The capacity of a flow cytometer to measure fluorescence intensity may be helpful in deciding the specificity of staining when background fluorescence is present. Present methods use microscopic examination of primary specimen smears stained with an immunofluorescent conjugate. This is a tedious and time consuming method.

The small, free-living amoebas <u>Naegleria</u> and <u>Acanthamoeba</u> may be present in soil and water. Their numbers may be increased by natural or industrial heating of water. Swimmers in these bodies of water may be infected, resulting in meningoencephalitis, pneumonitis, and conjunctivitis. Culture and identification of amoebas are laborious and often not successful. Identification usually is based on morphology of the various life stages. An efficient system for detecting these amoebas is essential for research on their growth in the environment. Public health laboratories also could use such technology for examining the safety of waters for swimmers. Muldrow et al [78] used flow-cytometric analysis of river water samples labeled with immunofluorescent reagents and fluorescein-conjugated concanavalin A (ConA-F) to detect and identify these amoebas. These specific reagents allowed detection of <u>Naegleria fowleri,</u> the pathogen, and its distinction from nonpathogenic <u>Naegleria.</u> The capacity to distinguish and count large numbers of <u>Naegleria</u> rapidly enabled studies of generation time and competitive growth of two <u>Naegleria</u> species with a precision previously impossible. Standard mouse pathogenicity tests for <u>Naegleria fowleri</u> were corroborated by the more rapid and sensitive flow-cytometric method. Con A-F binds to <u>N. lovaniensis</u> in high concentration, permitting easy discrimination from other <u>Naegleria</u> species and from <u>Acanthamoeba.</u> The highly absorbed immunofluorescent reagents used in this study or specific monoclonal antibody combined with flow cytometry should be used to analyze waters in highly endemic regions. The sensitivity of this system would be most useful for studies of human infection.

XI. CLINICAL MICROBIOLOGY

Present clinical microbiology procedures usually require a culture period of at least 18-24 h before bacteria or yeasts are seen as a colonial growth. Direct examination of clinical specimens with detection and identification of microorganisms would reduce markedly the time required to establish the correct diagnosis and start appropriate therapy. Flow cytometry combined with some system for microbial identification, eg, DNA staining to yield percentage GC, immunofluorescence with monoclonal antibody, or DNA probe homology in situ, offers great promise for such a diagnostic system.

Van Dilla et al [59] were able to detect bacteria directly in the sediment from infected urine specimens, count the total number of bacteria per unit volume, and distinguish different kinds of bacteria present based on their staining by chromomycin A3 and Hoechst 33258 (see Section V). Common contaminants in urine, such as <u>Staphylococcus epidermis</u> and <u>Corynebacterium</u> sp., could be detected and distinguished from infecting bacteria, Escherichia coli, Streptococcus faecalis, and Staphylococcus saprophyticus. The relative numbers of each could be determined. This two-parameter analysis requires a dual laser beam flow cytometer. Processing of the urine specimen is simple and follows the sequence centrifugation, ethanol fixation of sediment, application of the two stains in buffer, and flow-cytometric analysis of whole staining solution and stained sediment. Fragmented leukocytes, whose nuclear fragments are also stained by these DNA stains, do not interfere because they are larger than bacteria and display very great fluorescence intensity.

In an extension of this work, antimicrobial susceptibility of bacteria in a urine sediment was measured by incubation with buffered broth culture medium containing varying concentrations of antimicrobial for 1-2 h. After the sediment in each tube was fixed and stained, the bacteria were counted and the number was compared with the control, which was not exposed to an antimicrobial agent. The minimal inhibitory concentration (MIC) of the antimicrobial agent was indicated when there was no increase in number of cells counted over the control.

Monsour et al [79] have been able to detect Escherichia coli directly in blood using flow cytometry. A growth period was not necessary provided sufficient bacteria were present in the initial specimen. Specimen preparation extended over 2 h and consisted of lysis of blood cells, staining with ethidium bromide in buffer, centrifugation onto a fluorinert FC-40 cushion, and resuspension of the sediment in saline. The sample was analyzed by a FACS IV flow cytometer using a 488-nm laser line. Analysis of forward scatter and the ratio of green to red fluorescence was necessary to discriminate autofluorescent particulates (green) from ethidium bromide-stained bacteria (red). The E. coli were detected at levels of 10-100 bacteria per milliliter when seeded in whole human blood, which contains greater than 10^9 eukaryotic cells per milliliter. The bacterial detection sensitivity achieved is 100 to 1000-fold greater than the detection sensitivity of most light or fluorescence visual microscopes. Numbers of bacteria that occur during bacteremia were detected without a growth period. Incubation of a blood culture for a few hours would further increase a remarkably sensitive analytical system.

XII. FUTURE APPLICATION OF FLOW CYTOMETRY TO MICROBIOLOGY

Flow cytometry has the potential for rapid detection and analysis of microorganisms based on measurement of either scattered light or fluorescence. A remarkable array of new fluorochromes with heightened fluorescent intensities is available. Some of these stains bind directly to specific chemical substances, or they can be conjugated to ligands, which bind to known receptors or to specific segments of molecules. Fluorogenic substrates, widely used for detecting enzymes produced by microorganisms, deserve study by flow cytometry. Fluorescently conjugated monoclonal antibodies are becoming available and have obvious use for the detection of antigens and for identification of microorganisms. The cytochemical detection of specific DNA or RNA sequences by hybridization with complementary sequences has been linked to fluorochrome labels by several methods [80,81]. Each of these technologies can be linked to rapid methods of flow cytometry and offers remarkable capacity for analysis, detection, and identification of microorganisms.

XIII. REFERENCES

1. Kamentsky, L. A.; Melamed, M. R.; Derman, H. Science, 1965, 150, 630.
2. Van Dilla, M. A.; Trujillo, T. T.; Mullaney, P. F.; Coulter, J. R. Science, 1969, 163, 1213.
3. Dittrich, W.; Gohde, W. Z. Naturforsch, 1969, 24b, 360.
3a. Kruth, H. S. Anal. Biochem., 1982, 123, 225.
4. Crissman, H. A.; Steinkamp, J. A. J. Cell Biol., 1973, 59, 766.
5. Shapiro, H. M. Cytometry, 1983, 3, 227.
6. Steinkamp, J. A.; Orlicky, D. A.; Crissman, H. A. J. Histochem. Cytochem., 1979, 27, 273.
7. Steinkamp, J. A.; Fulwyler, M. J.; Coulter, J. R.; Hiebert, R. D.; Horney, J. L.; Mullaney, P. F. Rev. Sci. Instrum., 1973, 44, 1301.
8. Horan, P. K.; Wheeless, L. L., Jr. Science, 1977, 198, 149.
9. Van Dilla, M. A.; Mendelsohn, M. L. In "Flow Cytometry and Sorting"; Melamed, M. R.; Mullaney, P. F.; Mendelsohn, M. L., eds.; John Wiley and Sons: New York, 1979, pp. 11-37.
10. Herzenberg, L.; Sweet, R. Sci. Am., 1975, 234, 108.
11. Gray, J. W.; Dean, P. N.; Mendelsohn, M. L. In "Flow Cytometry and Sorting"; Melamed, M. R.; Mullaney, P. F.; Mendelsohn, M. L., eds.; John Wiley and Sons: New York, 1979; pp. 383-407.
12. Kamentsky, L. A. Adv. Biol. Med. Phys., 1973, 14, 93.
13. Laerum, O. D.; Farsund, T. Cytometry, 1981, 2, 1.
14. Kute, T.; Linville, C.; Barrows, G. Cytometry, 1983, 4, 132.
15. Carrano, A. V.; Gray, J. W.; Langlois, R. G.; Burkhart-Schultz, K. J.; Van Dilla, M. A. Proc. Natl. Acad. Sci. U.S.A., 1979, 76, 1382.
16. Langlois, R.; Carrano, A.; Gray, J.; Van Dilla, M. Chromosoma, 1980, 77, 229.
17. Lebo, R. V. Cytometry, 1982, 3, 145.
18. Hoffman, R. A.; Kung, P. C.; Hansen, P.; Goldstein, G. Proc. Natl. Acad. Sci. U.S.A., 1980, 77, 4914.
19. Loken, M. R. J. Immunol. Methods, 1982, 50, 85.
20. Lanier, L. L.; Engleman, E. G.; Gatenby, P.; Babcock, G. F.; Warner, N. L.; Herzenberg, L. A. Immunol. Rev., 1983, 74, 147.
21. Bassoe, C.-F.; Laerum, O. D.; Solberg, C. O.; Haneberg, B. Proc. Soc. Exp. Biol. Med., 1983, 174, 182.
22. Svenson, S. B.; Kallenius, G. Infection, 1983, 11, 6.
23. Cram, L. S.; In "Second International Symposium on Rapid Methods

and Automation in Microbiology, Cambridge, England"; Johnston, H. H.; Newsom, S. W. B., eds.; Learned Information, Ltd.: Oxford, 1976; pp. 215-220.

24. Melamed, M. R.; Mullaney, P. F.; Mendelsohn, M. L., eds.; "Flow Cytometry and Sorting"; John Wiley and Sons: New York, 1979.

25. Brattain, M. G. In "Flow Cytometry and Sorting"; Melamed, M. R.; Mullaney, P. F.; Mendelsohn, M., eds.; John Wiley and Sons: New York, 1979; pp. 193-205.

26. Dolbeare, F. A.; Smith, R. E. In "Flow Cytometry and Sorting"; Melamed, M. R.: Mullaney, P. F.; Mendelsohn, M., eds.; John Wiley and Sons: New York, 1979; pp. 317-333.

27. Darzynkiewicz, Z. In "Flow Cytometry and Sorting"; Melamed, M. R.; Mullaney, P.; Mendelsohn, M., eds.; John Wiley and Sons: New York, 1979; pp. 283-316.

28. Kerker, M.; VanDilla, M. A.; Brunsting, A.; Kratohvil, J. P.; Hus, P.; Wang, D. S.; Gray, J. W.; Langlois, R. G. Cytometry, 1982. 3, 71.

29. Waggoner, A. S. Ann. Rev. Biophys. Bioeng., 1979, 8, 47.

30. Rotman, B.; Papermaster, B. W. Proc. Natl. Acad. Sci. U.S.A., 1966, 55, 134.

31. Widhohn, J. M. Stain Technol., 1972, 47, 189.

32. Visser, J. W. M. Acta Pathol. Microbiol. Scand., 1981, Suppl. 274, 86.

33. Van Epps, D. E. J. Reticulo-endothel. Soc., 1983, 34, 113.

34. Douglas, R. H.; Ballou, C. E. J. Biol. Chem., 1980, 255, 5979.

35. Visser, J. W. M.; Van Den Engh, G. J. In "Immunofluorescence Technology"; Wick, G.; Traill, K. N.; Schauenstein, K., eds.; Elsevier Biomedical Press: Amsterdam, 1982; pp. 95-128.

36. Heitzman, H.; Richards, F. M. Proc. Natl. Acad. Sci. U.S.A., 1974, 71, 3537.

37. Latt, S. A. In "Flow Cytometry and Sorting"; Melamed, M. R.; Mullaney, P.; Mendelsohn, M., eds.; John Wiley and Sons: New York, 1979; pp. 263-284.

38. Crissman, H. A.; Stevenson, A. P.; Kissane, R. J.; Tobey, R. A. In "Flow Cytometry and Sorting"; Melamed, M. R.; Mullaney, P.; Mendelsohn, M., eds.; John Wiley and Sons: New York, 1979; pp. 243-261.

39. Nairn, R. C. "Fluorescent Protein Tracing", Fourth ed.; Churchill Livingstone: Edinburgh, 1976.

40. Crissman, H. A.; Steinkamp, J. A. Cytometry, 1982, 3, 84.

41. Latt, S. A.; Wohlleb, J. C. Chromosoma, 1975, 52, 297.

42. Mueller, W.; Gautier, F. Eur. J. Biochem., 1975, 54, 385.

43. Oi, V.; Glazer, A. N.; Stryer, L. J. Cell Biol., 1982, 93, 981.

44. Carlson, J.; Drevin, H.; Axen, R. Biochem. J., 1978, 173, 723.

45. Titus, J. A.; Haugland, R. P.; Sharrow, S. O.; Segal, D. M. J. Immunol. Meth., 1982, 50, 193.

46. Paau, A. S.; Cowles, J. R.; Oro, J. A. Can. J. Microbiol., 1977, 23, 1165.

47. Paau, A. S.; Lee, D.; Cowles, J. R. J. Bacteriol., 1977, 129, 1156.

48. Bailey, J. E.; Fazel-Madjlessi, J.; McQuitty, D. N.; Lee, L. Y.; Allred, J. C.; Oro, J. A. Science, 1977, 198, 1175.

49. Hutter, K.-J.; Eipel, H. E. Eur. J. Appl. Microbiol. Biotechnol., 1979, 6, 223.
50. Olaiya, A. F.; Sogin, S. J. J. Bacteriol., 1979, 140, 1043.
51. Gilbert, M.; McQuitty, D. N.; Bailey, J. E. Appl. Environ. Microbiol., 1978, 36, 615.
52. Agar, D. W.; Bailey, J. E. Cytometry, 1982, 3, 123.
53. Steen, H. B.; Boye, E. Cytometry, 1980, 1, 32.
54. Steen, H. B.; Boye, E. J. Bacteriol., 1981, 145, 1091.
55. Steen, H. B.; Boye, E.; Skarstadt, K.; Bloom, B.; Godal, T.; Mustafa, S. Cytometry, 1982, 2, 249.
56. Boye, E.; Steen, H. B.; Skarstadt, K. J. Gen. Microbiol., 1983, 129, 973.
57. Steen, H. B. Cytometry, 1980, 1, 26.
58. Martinez, O. V.; Gratzner, H. G.; Malinin, T. I.; Ingram, M. Cytometry, 1982, 3, 129.
59. Van Dilla, M. A.; Langlois, R. G.; Pinkel, D.; Yajko, D.; Hadley, W. K. Science, 1983, 220, 620.
60. Langlois, R.; Carrano, A.; Gray, J.; Van Dilla, M. Chromosoma, 1980, 77, 229.
61. Langlois, R.; Van Dilla, M.; Yajko, D.; Hadley, W. K. Personal communication, 1983.
62. Ingram, M.; Cleary, T. J.; Price, B. J.; Price, R. L., III; Castro, A. Cytometry, 1983, 3, 134.
63. Phillips, A. P.; Martin, K. L. Cytometry, 1983, 4, 123.
64. Waldman, F.; Schachter, J.; Fulwyler, M. J.; Hadley, W. K. Personal communication, 1984.
65. Hercher, M.; Mueller, W.; Shapiro, H. M. J. Histochem. Cytochem., 1979, 27, 350.
66. Notter, M. F.; Leary, J. F.; Balduzzi, P. C. J. Virol., 1982, 41, 958.
67. Murray, J. D.; Berger, M. L.; Taylor, I. W. J. Gen. Virol., 1981, 57, 221.
68. Gershey, E. L. Cytometry, 1980, 1, 49.
69. Dunn, J.; Spizizen, J.; Meinke, W. J. Histochem. Cytochem., 1978, 26, 391.
70. Aurelian, L. Anal. Quant. Cytol., 1979, 1, 89.
71. Leary, J. F.; Notter, M. F. D.; Todd, P. J. Histochem. Cytochem., 1976, 24, 1249.
72. Jacobberger, J. W.; Horan, P. K.; Hare, J. D. Cytometry, 1983, 4, 228.
73. Howard, R. J.; Battye, F. L.; Mitchell, G. F. J. Histochem. Cytochem., 1979, 27, 803.
74. Whaun, J. M.; Rittershaus, C.; Ip, S. H. C. Cytometry, 1983, 4, 117.
75. Howard, R. J.; Rodwell, B. J. Exp. Parasitol., 1979, 48, 421.
76. Trask, B. J.; Van Den Engh, G. J.; Elgershuizen, J. H. B. W. Cytometry, 1982, 2, 258.
77. Bonaly, J.; Mestre, J. C. Cytometry, 1981, 2, 35.
78. Muldrow, L. L.; Tyndall, R. L.; Fliermans, C. B. Appl. Environ. Microbiol., 1982, 44, 1258.

79. Monsour, J. D.; Robson, J. A.; Arndt, C. W.; Schulte, T. H. Abstract C 327, in "Abstracts of the Annual Meeting of the American Society for Microbiology", 1984.
80. Bauman, J. G. J.; Wiegant, J.; Van Duijn, P. J. Histochem. Cytochem., 1981, 29, 238.
81. Brigati, D. J.; Myerson, D.; Leary, J. J.; Spalholz, B.; Travis, S. Z.; Fong, C. K. Y.; Hsuing, G. D.; Ward, D. C. Virology, 1983, 126, 32.

4. THE IDENTIFICATION, INTERACTIONS AND STRUCTURE

OF VIRUSES BY RAMAN SPECTROSCOPY

K. A. Hartman and G. J. Thomas, Jr.

I. <u>INTRODUCTION</u>

The study of a virus by biochemical and biophysical methods begins with the isolation of the pure virus (with concomitant measures of viability). It proceeds to the determination of the molecular composition (number of molecules of DNA or RNA, number and kinds of proteins in the capsid, and the existence and composition of envelopes, tail structures, or proteins of special function) and continues with the investigation of the life cycle (penetration of the cell, disassembly, biochemical reproduction of constituent molecules, reassembly, and dispersion from the cell). A complete study culminates in the determination of the molecular organization of the components, including their intramolecular structures and intermolecular interactions.

A variety of physical methods has been used to investigate the molecular structures of viral particles, including X-ray diffraction, neutron diffraction, circular dichroism, absorption, fluorescence and NMR spectroscopy, potentiometric titration, light scattering, electron microscopy, and, most recently, laser Raman spectroscopy. Although X-ray diffraction holds the promise of providing the coordinates of all atoms (except hydrogen) in the virus particle, this task is exceedingly difficult and has yet to be accomplished in totality for any virus. Therefore, other physical methods, which may address narrower objectives or answer very specific questions in considerable detail, are important adjuncts to diffraction methods. Laser Raman spectroscopy has been one of the most recently developed of these methods and it shows considerable promise as an effective and powerful tool in virus structure research.

Reviews of physical and chemical methods as applied to virology are numerous [1-3]. However, no comprehensive review of applications of Raman spectroscopy in virus research has been given previously, even though several treatises on biological applications of Raman spectroscopy [4-8] necessarily cover some of the constituent molecules of viruses. Accordingly, this chapter provides a discussion addressed to virologists and cell biologists with reference to more specialized reviews as appropriate.

A Raman spectrum is obtained by placing the sample in a beam of monochromatic radiation and measuring the intensity of the light scattered at longer wavelengths (or smaller values of frequency). Modern instruments plot the intensity of scattered light versus the difference between its frequency and the frequency of the exciting line. This frequency difference is often called the "Raman shift" or, more simply, the frequency or wavenumber (in cm^{-1} units) of the Raman line. Each Raman frequency corresponds to the frequency of the vibration of a molecular subgroup and is influenced by the nature of the covalent bonds, the geometrical structure of the group of atoms, and inter and intramolecular interactions, such as hydrogen bonding. The spectral intensity of a Raman line, like its frequency, also may be influenced by molecular geometry and secondary interactions, so that the Raman spectrum of a protein, nucleic acid, or virus contains much detailed information in both the positions of the peaks and their relative intensities [6,7].

As explained in Section II, the Raman spectrum of a virus may provide the following kinds of information: (1) the primary structure of molecular constituents (base and amino acid compositions); (2) the nature of the secondary structures of the nucleic acid (A, B, or other helical forms) and proteins (α -helix, β -sheet, and irregular conformations); (3) the state of protonation of certain DNA or RNA bases; (4) the extent of base stacking and pairing; (5) the environment or orientation of side chains of certain amino acid residues; and (6) changes in any of the above as a function of chemical (eg, ionic) composition or physical (eg, thermal) perturbations. It is often particularly fruitful to compare the structural details of viral components with those of the complete virus particle to reveal interactions between component molecules [8].

Raman spectroscopy began in the late 1920s with its discovery by Sir C. V. Raman. Original instruments were photographic, used mercury arcs for excitation, and required huge quantities of sample by modern biochemical standards. Even photoelectric recording instruments, developed after 1945, required samples so large as to exclude all but the most simple and readily available of biological molecules. Therefore, the pioneering work by John Edsall and his associates in the prelaser era was limited for the most part to study of amino acids and peptide-related organic compounds.

This situation changed significantly with the availability in the early 1960s of continuous-wave lasers of significant power. The laser provided an intense and coherent beam of monochromatic radiation that could be focused easily on a small volume of sample to yield Raman spectra of quality far superior to that possible with a mercury-arc instrument. Laser sources, as well as better monochromators, detectors, and electronics, produced renewed interest in Raman spectroscopy. By the late 1960s spectra of 200 µg of RNA could be recorded and the structures and interactions of nucleic acids and proteins were given vigorous attention [7]. The first Raman spectrum of a virus was published in 1973 [9]. In the last decade, many other viral species have been studied and methods of analysis have been perfected. However, the quantity of sample required for Raman spectroscopy continues to play a limiting role in virus applications. Present technology effectively limits the subjects of study to plant and bacterial viruses. Animal viruses, which are generally more limited in availability, pose a formidable challenge for detailed study by Raman spectroscopy.

This chapter explains the principles and methods of Raman spectroscopy and how they are used to measure the structures of and interactions between the molecular components of viruses. The results obtained for several representative viruses are reviewed to see whether current knowledge allows us to distinguish between different classes of viruses by comparing Raman spectra. Finally, new methods are reviewed that promise to increase the speed and sensitivity of Raman spectroscopy. At this point it is appropriate to note that the format for presentation of Raman spectra is currently undergoing some change in most laboratories as computerization of Raman spectrometers is implemented. The reader will observe, for example, that in some of the spectra presented below, the energy or wavenumber axis (x axis) is increasing from right to left, typical of the precomputer era. On the other hand, in more recently recorded data the abscissa displays wavenumber increasing from left to right, which appears to be the format of choice in laboratories using spectrometers under computer control.

II. THE THEORY AND PRACTICE OF RAMAN SPECTROSCOPY

A. General Characteristics of Raman Spectra

Raman scattering spectra, like infrared absorption spectra, originate from the exchange of energy between photons and vibrational or rotational motions in molecules. In biological applications only condensed phases (solutions, liquids, crystals, etc) are encountered ordinarily, and consequently molecular rotations can be ignored. Therefore, the Raman spectrum of a biological molecule is in effect a vibrational spectrum.

The Raman spectrum arises when quanta of vibrational energy are transferred between the vibrating molecules and the (laser) beam of photons incident upon the sample. The spectrum is contained in the inelastically scattered radiation. Because the transfer mechanism (Stokes transition) is intrinsically very inefficient, the intensity and frequency of the scattered light must be monitored accurately and precisely. Accordingly, a Raman spectrophotometer must have design features that facilitate the rejection of both elastically scattered radiation (Rayleigh scattering) and stray light.

As in the case of infrared spectroscopy, Raman spectroscopy profits as an analytical technique from the fact that the molecular vibrational spectrum is a highly characteristic and structurally sensitive fingerprint of the molecule. An added virtue of Raman over infrared spectroscopy for biological applications is the low level of interference that arises from liquid H_2O. Unlike the infrared spectrum, the Raman spectrum of solvent H_2O is fairly weak when compared with that of a typical solute biomolecule. In other words, the Raman scattering efficiency of an amino acid (protein constituent) or nucleotide (nucleic acid constitutent) is substantially greater than that of H_2O throughout the spectral range of interest to a biochemist.

Figure 4-1 shows the Raman spectra of liquid H_2O and D_2O throughout the vibrational spectral range (300-4000 cm^{-1}) normally measured in biological applications. Except for broad bands (or "lines") caused by stretching modes, little solvent interference is encountered. By using both H_2O and D_2O as a complementary solvent pair, virtually all of the molecular vibrational spectrum of a solute biomolecule can be obtained.

B. Raman Spectra Reveal Molecular Properties

Molecular vibrations. The modes of vibration of a polyatomic molecule can be reasonably well approximated using a model of nuclear point masses connected by springs. In this approximation, the atoms oscillate about their equilibrium positions with restoring forces that are proportional to their displacement from the equilibrium molecular configuration (harmonic oscillation). The restoring forces will in general be determined by the stiffness and polarity of the covalent bonds and by nonbonded interactions as well, including dipolar and van der Waals interactions. Therefore, the energetics of the molecular vibrations are influenced by both the intramolecular chemical bonds and the molecular environment. In the harmonic oscillator model, the energy of the vibrating unit is related also to the inverse square root of the vibrating masses. A given vibrational mode of a molecule therefore can be expected to exchange a quantity of energy (quantum) with incident electromagnetic radiation that is determined by (1) the masses of the atoms in motion, (2) the nature of the covalent bonds between the atoms, and (3) other intermolecular and intramolecular interactions in which the atoms participate, such as hydrogen bonding, hydrophobic bonding, electrostatic bonding, and the like.

The preceding considerations relate to each and every independent mode of vibration (normal mode) of the molecule. For a polyatomic molecule of N atoms, as many as $3N - 6$ normal modes are possible. Fortunately, not each of these leads to Raman scattering at a unique position in the spectrum, ie, to a Raman line; otherwise the Raman spectrum of even a relatively small biological molecule, such as chicken egg-white lysozyme, would be cluttered with thousands of individual Raman lines. The Raman spectrum in fact is simplified by the following constraints. First, in a polymer, each repeat unit of identical structure generates, to a first approximation, the same Raman scattering. If the repeat unit contains n atoms, then no more than $3n-6$ Raman lines can arise from all such identical units. For example, in lysozyme, to a first approximation, all three tyrosine residues give Raman lines at the same positions in the spectrum. Therefore, the observed tyrosine Raman lines are the composite of contributions from residues 20, 23, and 53 in the sequence. Second, not all of the allowed $(3n-6)$ vibrational modes of a molecular subgroup can interact with electromagnetic radiation to yield Raman scattering. On the contrary, Raman scattering occurs if and only if the

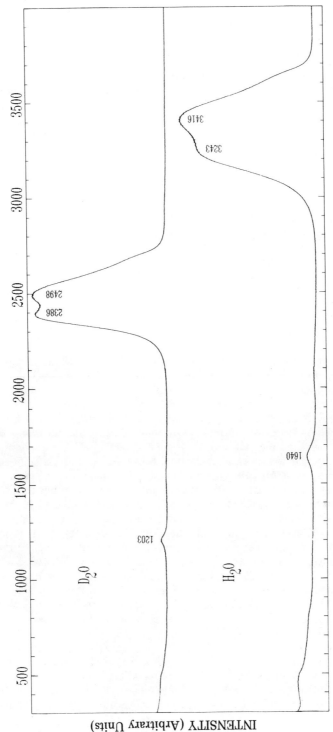

Figure 4-1. Raman spectra of normal liquid water (H_2O) and heavy water (D_2O) in the range 300-4000 cm^{-1}. Note that cm^{-1} values increase from left to right along the abscissa, in accord with current practice (cf. Figures 4-10 through 4-16) but opposite to the format of earlier work (cf.Figures 4-4 through 4-9).

normal mode in question involves a change of polarizability of the molecule as the vibration occurs. Further, the greater the change in polarizability, the greater is the intensity of the corresponding Raman scattering line. This consideration greatly reduces the effective contribution to the Raman spectrum from each kind of repeat unit in the macromolecule. In the case of tyrosine residues of lysozyme, for example, only two or three of the theoretically allowed 30 normal modes (n = 12 is assigned for the tyrosine side group) involve a sufficiently large polarizability change to permit an intense Raman line in the protein spectrum. In fact, the pair of intense Raman lines of tyrosine (usually ca. 830 and 850 cm^{-1}; see Table 4-2 and Section III, A) are of great usefulness in the analysis of protein structure [10]. In addition, there are a few moderately intense and a few very weak Raman lines that originate from tyrosine; however, these are of limited usefulness. Similar arguments apply to each of the 20 amino acid side groups and also to the peptide backbone. These considerations account for the occurrence of only about 40-50 lines in the Raman spectrum of lysozyme.

The same constraints applied to a DNA molecule account for the reduction of the theoretically allowed number of lines to about 10 lines from each of the bases (A, T, G, C) and another half-dozen or so from the sugar-phosphate backbone. Properly analyzed, the Raman spectrum of a protein or nucleic acid can yield structural information about many of the constituent subgroups and the skeletal backbone [8,11,12].

Intensities. At this point, it is appropriate to consider the molecular bonding arrangements that generate vibrational modes of high polarizability change in biological molecules. Stated otherwise, what kinds of vibrations of molecular subgroups of proteins, nucleic acids, lipids, etc, lead to intense Raman lines? Generally, vibrations of heavy atoms and of large groups of atoms, as well as vibrations of multiply bonded atoms, lead to intense Raman scattering. For example, vibrations of C=C, C=O, C=N double-bonds, and other multiply bonded atoms, especially when these groups are conjugated in aromatic rings, yield the most intense Raman lines in spectra of nucleic acids and proteins. Also intense are the Raman lines from stretching vibrations of very heavy atoms, such as S-S groups of cystine in proteins and Fe-N groups of iron-ligand centers in metalloproteins. Conversely, vibrations of H atoms, such as in C-H bond stretching and bending motions, are predictably very weak. When very many CH groups are present, nevertheless, their cumulative Raman intensity can rival that of heavy atom vibrations. Raman spectra of lipids are dominated by lines that arise from vibrational modes of the relatively heavy fatty acid ester and phosphoester groups, and by concerted motions of the hydrocarbon chains as a whole. The large number of methylene groups in the chains assures that some intense Raman lines also arise from them despite the inherent weakness of the normal modes of isolated CH_2 groups.

Finally, certain distinctions between Raman spectra and other types of optical spectra familiar to biochemists can be noted. Unlike infrared, and ultraviolet absorption spectra (which measure the intensity of the light transmitted by the sample) the intensity of Raman scattering cannot be expressed by using absorbance or percentage transmittance. A precise quantitative treatment of the intensity of Raman scattering would require (among other things) a consideration of the scattering distribution over the full solid angle (4π stereradians). This is not feasible experimentally. The intensity of Raman scattering, ie, the ordinate scale of the spectrum, is expressed in arbitrary units, such as the number of photon counts recorded by a detector. Quantitative inferences can be made by comparison of the Raman line intensities in an unknown sample with those in a known standard. An example is the comparison of intensities of the tyrosine doublet in lysozyme with corresponding intensities in model compounds of known structure (eg, peptides that contain tyrosine residues of known geometrical orientation and environment).

Also, in contrast to infrared absorption intensities, Raman scattering intensities are relatively insensitive to changes in the polarity (dipole moment) of molecular subgroups. Therefore, if a potential hydrogen bonding group, such as the phenolic OH group of tyrosine, is transferred from a state of weak hydrogen bonding to one of strong hydrogen bonding, this environmental change alone cannot be expected to generate significant change in the intensity of the OH vibrational modes of tyrosine in the Raman spectrum. The intensities of key Raman lines of tyrosine residues are indeed altered significantly by changes in the hydrogen bonding environment of the p-hydroxyl group (see Table 4-2 and Section III, A), not because of the direct effects of hydrogen bonding per se, but because of the phenomenon of Fermi resonance, which exerts a much more profound influence on the intensities of the Raman lines in question [10].

Frequencies. The Raman spectrum of a molecule is a graph which shows the intensity of the inelastically scattered photons as a function of the wave number of the photons. In the usual Raman spectrometer, photons of known wavelength (λ) or wavenumber (σ, defined as $1/\lambda$ and expressed in cm^{-1} units), generated by a continuous-wave (CW) laser, are focused on the sample. The number of scattered photons (with wavenumber less than the wavenumber of the incident beam, $\sigma < \sigma_0$) are then recorded. The plot of the intensity in arbitrary units of this inelastically scattered light versus the wavenumber constitutes the Raman spectrum. The wavenumber and corresponding wavelength for exciting lines from some CW lasers commonly employed in biochemical studies are listed in Table 4-1.

For consistency with infrared vibrational spectroscopy, and in accord with the mechanism of Raman scattering, the abscissa of the spectrum records the difference in wavenumber between incident and scattered photons ($\sigma_0 - \sigma$), rather than the absolute wavenumber. For example, the spectrum of H_2O in Figure 4-1 displays its most intense Raman scattering at 3416 which means that the scattered photons have a wavenumber which is 3416 cm^{-1} less than the laser beam used for excitation. This corresponds to the transfer of quanta of vibrational energy equivalent to 3416 cm^{-1}, from the incident laser beam to the H_2O molecules. Here the energy is utilized by H_2O to excite OH stretching vibrations. This particular spectrum was obtained by irradiation of

Table 4-1. The Wavelengths and Wavenumbers of Radiation Available
from Continuous-Wave Lasers

Wavelength (nm)	Wavenumber (cm^{-1})	Source
647.09	15,454	Kr
632.82	15,802	He-Ne
530.87	18,837	Kr
514.53	19,435	Ar
487.99	20,492	Ar
468.04	21,366	Kr
413.13	24,206	Kr
406.74	24,586	Kr
363.79	27,489	Ar
351.11	28,481	Ar
350.74	28,511	Kr

the H_2O sample with the 514.53 nm line of the argon ion laser. However, the same spectrum would be obtained by use of any of the laser excitation wavelengths of Table 4-1 because the difference in wavenumber between incident and scattered photon, rather than the absolute wavenumber of the photon, is the parameter of interest in Raman spectroscopy.

Nonresonance Raman spectra. The Raman transition illustrated by the solid arrow (path a in Figure 4-2) represents an ordinary or nonresonance Raman transition. The laser photon is of insufficient energy (or wavenumber, σ_0) to promote the molecule from its ground electronic state (g) to one of its excited electronic states (e). Hence the photon is scattered and not absorbed. After scattering has occured, the target molecule is left in a higher vibrational energy level. The molecule has thereby gained energy from the incident photon. Simultaneously the scattered photon is diminished in energy and therefore in wavenumber (σ_v). A sensitive monitoring of the intensity of light scattered as a function of wavenumber reveals an increase (ie, a Raman line) at $\sigma_0 - \sigma_v$, corresponding to those quanta of energy transferred to the molecule.

Resonance Raman spectra. If the laser frequency is deliberately chosen (σ_R) to match the gap between ground and excited electronic states, then normal Raman scattering cannot occur because the photons are both absorbed and scattered, leading to the resonance Raman effect. This process is depicted by the open arrow (path b in Figure 4-2). The process of resonance Raman scattering is several orders of magnitude more efficient than that of nonresonance Raman scattering. Therefore, resonance Raman spectra can be obtained on much more dilute systems than required in the normal Raman effect. However, resonance Raman transitions are much more selective. Only normal

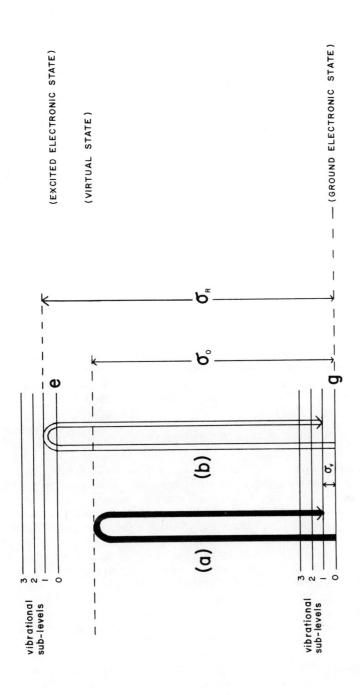

Figure 4-2. Energy level diagram illustrating Raman spectroscopic transitions: (a) Ordinary or nonresonance Raman transition, (b) resonance Raman transition. In a polyatomic molecule of N atoms, $3N-6$ such diagrams can be constructed, one for each degree (v_i) of vibrational freedom.

modes of vibration of the immediate chromophore (ie, the part of the molecule in which the electronic excitation takes place) can be excited by the resonance Raman effect. More specifically, only vibrations of the substructure that undergoes a change of geometry in the excited state are revealed.

Those metalloproteins that absorb radiation in the 500-550 nm range, are suitable targets for excitation of resonance Raman spectra. The resonance Raman spectrum of hemoglobin, for example, contains selected lines of the iron-porphyrin chromophore. On the other hand, proteins and nucleic acids, which do not absorb visible radiation, cannot be subjected to resonance Raman study with use of visible wavelength lasers (Table 4-1). For such systems, in which the lowest energy electronic transitions require ultraviolet excitation, the resonance Raman spectrum would require use of an ultraviolet laser.

The choice between nonresonance and resonance Raman spectroscopy therefore usually is dictated by the nature of the sample, the available instrumentation, and the kind of information sought.

C. Instrumentation

The detailed design of Raman instrumentation varies from one laboratory to another depending on the kind of sample under investigation and the specific goals of the investigator. The apparatus shown schematically in Figure 4-3, however, depicts the essential features of many Raman spectrometers and is suitable for both nonresonance and resonance Raman applications. The reader is referred to Carey's monograph [4] for further details.

D. Sample-Handling Techniques

General requirements. In order to obtain a satisfactory Raman spectrum, interaction of the exciting radiation with the sample by such other mechanisms as absorption (including resonance Raman processes), fluorescence, and Tyndall scattering must be minimized. Absorption is eliminated by choice of a wavelength for exciting the Raman spectrum (Table 4-1) which is removed from any absorption bands of the sample. Fluorescence often can be controlled by appropriate choice of the excitation wavelength. Frequently, the fluorescence observed in Raman spectra of biological materials is not from the biomolecules themselves but from organic contaminants. Such contaminants usually are "burned up" after prolonged exposure of the sample to the laser beam. In other cases the use of time-discrimination techniques or the addition of a "fluorescence quencher", such as potassium iodide, may eliminate the problem.

Tyndall scattering, sometimes encountered with solutions, results from the presence in the sample cell of air bubbles or suspended particles, such as dust, colloids, or other undissolved matter, with particle size comparable to or

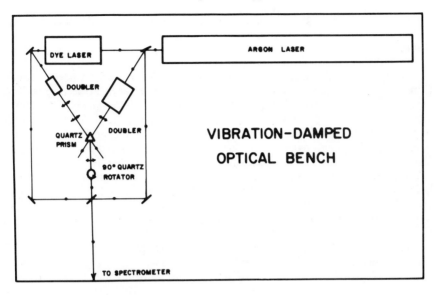

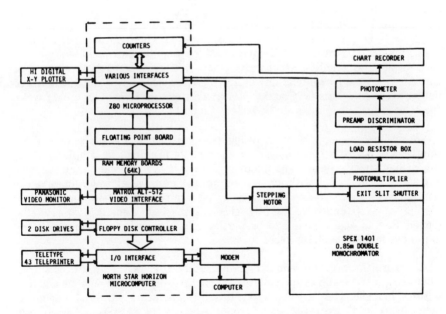

Figure 4-3. Schematic diagrams of computerized Raman spectrometer and laser optical bench. Reproduced with permission from Prog. Clin. Biol. Res. [34].

greater than that of the excitation wavelength. This type of light scattering detracts from the quality of the Raman scattering spectrum and is mostly eliminated by high-speed centrifugation or micropore filtration of the sample.

Solids. Modern laser Raman spectroscopy permits the routine examination of solids of many types, including single crystals, crystallites or powders, amorphous solids, and polymer films and fibers. The sample-handling details are often dictated by the specific information sought from the spectra as well as by sample morphology. Generally, the specimen is affixed to a goniometer or other positioning device and mounted directly in the laser beam. Samples also may be contained in glass or quartz capillary cells within which the relative humidity can be controlled precisely. The latter technique is particularly well suited to the study of fibrous materials, such as DNA or filamentous viruses. In the case of most solid samples, special care must be taken to prevent overheating or photodecomposition of the solid by the laser beam.

Aqueous solutions. One of the major advantages of Raman over IR spectroscopy for the study of biological materials is that aqueous solutions may be investigated with little or no solvent interference. Raman spectra of aqueous solutions of low viscosity are obtained with relative ease. Dilute solutions of biopolymers, or solutions of low molecular weight materials that are not aggregated, are investigated in the same manner as are solutions of inorganic or organic chemical materials. The concentration of solute required to obtain a satisfactory spectrum depends on the intrinsic intensity of Raman lines associated with the vibrational transitions being observed. For example, vibrations of aromatic rings and of linked heavy atoms (such as P—O, C—S, and S—S give prominent Raman lines when solute concentrations are as low as 0.02 M (approximately 0.3% by weight) and virtually noise-free spectra are obtained when solute concentrations are 0.2 M (3%). For carbohydrates and proteins, however, vibrations of singly bonded groups of lighter atoms (such as C—C and C—O) give prominent Raman lines only when the solute concentration is about 10% by weight.

Biopolymers, aggregated monomers, and other materials that form solutions of high viscosity may pose formidable problems in sample handling for Raman spectroscopy. One major difficulty is that of loading the cell with a viscous fluid or gel without loss of optical homogeneity. Ordinarily, the sample can be transferred by syringe, micropipet or air suction into the capillary cell (Kimax No. 34507 glass capillaries). Centrifugation of the loaded Raman cell is found to be particularly helpful in overcoming optical inhomogeneties in such samples.

A clear solution is not appreciably heated by the laser beam. For example, when 250 mW of 488.0 nm radiation is focused on a homogeneous aqueous solution in a glass capillary cell, the solution temperature will usually be in the range 30 to 35°C. However, when suspended particles are present, severe overheating of the sample may develop, which precludes the possibility of obtaining a satisfactory Raman spectrum. Devices for careful control of the sample temperature have been described [6].

Because Raman spectrophotometers usually are operated in a single-beam mode, spectral interference from a solvent is compensated by recording a spectrum of the pure solvent separately and then subtracting its spectrum from that of the solution, either manually or by computer methods. For quantitative applications, it is necessary to make use of an internal standard or to normalize all Raman intensities to a single Raman line of reliably constant intensity.

Polarized scattering and depolarization ratios. As in the case of polarized IR absorption, polarized Raman scattering can provide, in principle, an increase in the informational content of the spectra. For the special case of an oriented helical biopolymer, the theoretical aspects of polarized Raman scattering have been treated in detail and the angular dependence of Raman scattering from randomly oriented molecules has also been considered [4]. In practice, however, it is considerably more difficult to obtain polarized Raman scattering spectra than polarized IR absorption spectra. Experimental obstacles usually encountered are (1) the loss of unidirectional orientation of the molecules in the sample, possibly associated with the effects of laser illumination itself, and (2) the inability of conventional collection optics to discriminate the Raman scattering in a given direction (or plane) from the light scattered in other directions. Nevertheless, some recent progress has been made in overcoming these difficulties, and polarized Raman scattering from several helical biopolymers has proved to be structurally more informative than the corresponding nonpolarized Raman spectra.

Microsampling techniques. Virtually all laser Raman spectroscopy of solid and liquid materials is carried out on microliter or microgram quantities of sample. In the applications cited below solutions are investigated routinely in capillary cells requiring no more than a few microliters of sample, and less than milligram quantities of the solids are easily positioned in the focus of the laser beam. In the experience of the authors, the quantities of biological materials that are required for Raman spectroscopy are usually comparable to and frequently smaller than those required for IR spectroscopy.

Methods for obtaining satisfactory Raman spectra on nanogram and nanoliter quantities of sample have been described also [4].

E. A Typical Protocol

Let us assume that we wish to prepare a sample of the double-stranded DNA virus, P22, for analysis by Raman spectroscopy. The first requirement is authentication of the viability of the sample as infectious P22 phage particles. This can be accomplished by plaque assays that measure the number of viable phage in the sample. This number may be compared with the total number of phage particles present as determined from UV absorbance and known extinction coefficients. The ability of the phage particles to kill the host cells also may be measured by a cell survival assay. Examination of the phage particles with an electron microscope shows major structural defects but usually cannot determine viability. In a typical sample used for Raman spectroscopy, many phage particles will be nonviable but only a small fraction will have serious structural damage. Therefore the observed spectra accurately measure the structure and interactions found in viable virus particles.

The sample of phage, typically maintained at relatively low concentration (say 0.1 mg/mL) in tris pH7 buffer by the biochemist, requires concentration by about 300-fold (to 30 mg/mL or 3% by weight) for Raman spectroscopy. A convenient method for concentrating the phage is by pelleting in the ultracentrifuge. The pellet is resuspended, if necessary, in a new buffer devoid of interfering impurities and then repelleted. The final pellet can then be transferred by syringe to a Raman capillary cell. Some trial and error is required in order to arrive at the correct speed and time of ultracentrifugation that produces a pellet of the desired phage density.

Interpretation of data.   Spectroscopic data from complex molecules usually are interpreted by analogy with data from simpler model compounds of known structure or conformation. In this regard Raman spectroscopy is very similar to other types of molecular spectroscopy familiar to biochemists, such as circular dichroism (CD) and nuclear magnetic resonance (NMR) spectroscopy.

In past work, the Raman spectra of proteins were compared with spectra of simple polypeptides and amino acids of known structure. This allowed the assignment of most lines in the protein Raman spectra to vibrational modes of the amino acid side chains and the peptide main chain [13,14]. Analysis of Raman data from proteins of known structure (determined by X-ray crystallography) has provided additional correlations of spectral intensities and wavenumbers with specific protein conformations, such as α-helix, β-sheet, and nonrepetitive or irregular conformation [15]. Environments and interactions of selected side chains have been analyzed using similar procedures. Table 4-2 gives the wavenumbers for several such protein groups and lists the kind of structure information that can be inferred from the data.

In the case of nucleic acids, the characteristic Raman frequencies of nucleotides and polynucleotides have provided a basis for interpretation of spectra of RNA and DNA [11,12,16,18]. As DNA fibers and crystals of known conformation have become available, the data base has expanded [19,20] to permit characteristic Raman wavenumbers and intensities of specific nucleic acid structures to be identified and exploited for analysis of viruses [21,22]. Some of the important DNA and RNA correlations are listed in Table 4-3.

To return to the example of the preceding section (bacteriophage P22) the Raman spectra of the scaffolding protein, the procapsid, the empty procapside, the empty capsid, the tail-spike protein, the mature virus particle and its extracted DNA, all of which are shown in Figure 4-4 may be compared with Tables 4-2 and 4-3 to reveal the following: (1) the viral genome adopts the classical B-DNA structure within the protein shell of the capsid,

Table 4-2. Information Obtained from Some Selected Group Vibrations
in Proteins

Group	Wavenumber (cm^{-1})	Information
Aspartic and glutamic acids (COOH)	1715	Existence of protonated carboxyl groups
Amide I	1630-1680	Fractions of α-helix, β-sheet, and irregular structures
Amide III	1220-1290	Same as amide I
Cysteine SH	2550-2600	Existence and environment of S-H group
Cysteine SD	1850-1900	Existence and environment of S-D group
Tryptophan	760, 875, and 1360	Environment of the tryptophan ring
Phenylalanine	1005	Existence of phenylalanine
Tyrosine	830 and 850	Environment of the phenolic group
Methionine and cysteine C-S	630-750	Existence and conformation of CCSC and CCSS linkages
Cysteine S-S	530-545	Existence and conformation of the disulfide bridge (CCSSCC)

(2) this B-DNA secondary structure is not distinguishable from that of isolated, protein-free, double-stranded P22 DNA, (3) the tail protein is richest in β-sheet conformation, (4) the scaffolding protein is richest in α-helix conformation, and (5) the major coat protein contains a balanced mixture of both α-helix and β-sheet conformations. This secondary structure is essentially the same in the procapsid, empty procapsid, empty capsid, and mature virus particles. More detailed analyses of the P22 Raman spectra, including identification of side-chain interactions that differ among various assembly states of the virion, have been described [23].

More detailed discussions of the methods of analysis of Raman spectra of viruses and viral components are presented in Section III, below, with applications to single-stranded RNA and DNA bacteriophages and RNA plant viruses.

Table 4-3. Information Obtained from Some Selected Group Vibrations in RNA and DNA

Group	Wavenumber (cm^{-1})	Information
Guanine, cytosine, thymine and uracil (C=O groups)	1600-1700	Kind and extent of hydrogen bonding of the bases
Uracil ring	780, 1235, 1300	Presence of U and extent of base stacking
Thymine (T) ring	660, 750, 790, 1240, 1370	Presence of T and extent of base stacking
Adenine (A) ring	720, 1300, 1340, 1485, 1575	Presence of A and extent of base stacking
Guanine (G) ring	1320, 1380, 1485, 1575	Presence of G and extent of base stacking
Guanine (G) ring	625, 670 or 680	Guanosine conformation (A, B, or Z secondary structure)
Cytosine (C) ring (neutral)	785, 1248, 1292, 1527	Presence of C and extent of base stacking
Cytosine (C) ring (protonated)	1253, 1542	Presence of protonated C
PO_2^- symmetric stretch	1100	Intensity standard; also indicative of some ionic interactions of phosphate group
C-O-P-O-C	790-795	Indicates unordered nucleic acid backbone
C-O-P-O-C (RNA)	810-815	Indicates A-RNA backbone
C-O-P-O-C (DNA)	748, 792, 810 1095	Indicate Z-DNA backbone
	784, 830, 1090	Indicate B-DNA backbone
	806 - 810, 1099	Indicate A-DNA backbone

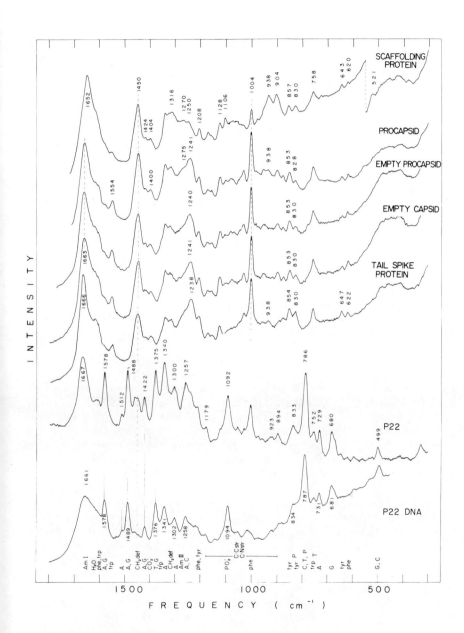

Figure 4-4. Raman spectra of P22 phage, precursor particles, and components. Reprinted with permission from Biochemistry 21, 250 - 265, copyright (1982) American Chemical Society.

III. RAMAN STUDIES OF VIRUSES AND THEIR COMPONENTS

A. Bacteriophage MS2

The first Raman spectrum of a virus was recorded [9] for the bacterio-phage R17 and work continued on the bacteriophage MS2 (a minor variant of R17), which was particularly well characterized by other methods [24]. The nearly spherical MS2 particle has a total molecular weight of 3.6×10^6, a diameter of 26 nm, and a sedimentation coefficient of 80 S and contains 70% protein and 30% RNA [1,2]. This virus is built of 180 identical coat-protein molecules (mol wt 1.1×10^6) and one unique protein molecule (mol wt 38,000), which is active in the infection process. When the Raman work was undertaken, the amino acid composition was known for the coat protein and the base composition was known for the RNA. The phage MS2 was also known to form special structures during heat denaturation (HD particles). Protein-free RNA and RNA-free capsid could be isolated using standard methods.

We next explain the method of analysis of the Raman spectrum of a virus to obtain information about the structures and interactions of the component molecules. Phage MS2 is used as an example [24]. The analysis begins with the spectrum of the RNA-free capsid and continues with the spectrum of protein-free RNA. The spectrum of the complete MS2 particle is then dis-cussed and compared with the spectra of its components.

Spectrum of the capsid. Spectra of RNA-free capsids in H_2O and D_2O solutions are shown in Figure 4-5. The C-H stretching region (about 2800 cm^{-1}) is best observed in curve B where the broad scattering from the O-H stretching bands of H_2O is absent. (Compare with Figure 4-1 which shows the spectra of pure H_2O and D_2O.) The aromatic C-H stretching line is seen near 3060 cm^{-1} and is weak in comparison with lines from aliphatic C-H groups near 2980, 2940, and 2880 cm^{-1}. No attempt has yet been made to construct a quantitative relationship between the intensities of these lines and the numbers of aromatic and aliphatic C-H groups in the protein.

The line from the S-H stretching motion of the two cysteine residues per coat-protein molecule is found near 2575 cm^{-1} (see curve A of Figure 4-4) and shifts to 1872 cm^{-1} for capsids in D_2O (curve B). This shift in frequency shows that the hydrogen atoms of the S-H groups exchange with deuterium in D_2O solution. We conclude that the S-H groups are accessible to solvent and are not buried in hydrophobic regions of the protein. Because no line from S-S is observed near 500 cm^{-1}, it is clear that the two cysteine residues are present in the reduced form. The weak line near 710 cm^{-1} in both spectra is most likely from the C-S bonds of cysteine and methionine residues.

The strong line from the amide I vibration (or amide I' for proteins in D_2O solution) is seen near 1660 cm^{-1} in both spectra of Figure 4-4. In curve A, the broad scattering band from liquid H_2O, which is centered near

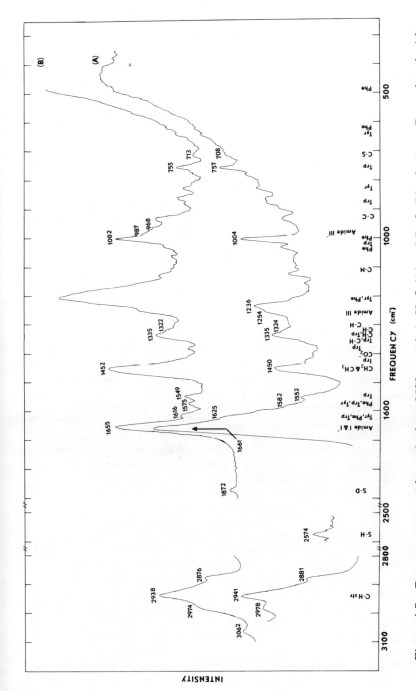

Figure 4-5. Raman spectra of bacteriophage MS2 capsids in H₂O (A) and D₂O (B) solutions. Reproduced with permission from J. Mol. Biol. [24].

1640 cm^{-1}, causes the amide I line to appear artificially intense, although the frequency of maximum scattering is not greatly altered. Curve B (in which the H_2O line is replaced by a D_2O line near 1210 cm^{-1}) shows the amide I' line without solvent interference.

The frequency value of the amide I' (or amide I) line suggests that most of the peptide groups in each coat-protein molecule participate in β-sheet or irregular structures with only a small fraction in α-helical regions. This is confirmed by the contours of the amide III region between 1200 and 1300 cm^{-1} in curve A, and of the amide III region between 900 and 1000 cm^{-1} in curve B. The amide I, I', III, and III' features predict only a small fraction of α helix in the coat protein of MS2.

Between 1630 and 1540 cm^{-1}, three lines from the aromatic amino acid residues are found in the spectrum of the capsid of MS2. These are not necessarily distinct lines and may arise from two or more different amino acid residues [eg, the 1625 cm^{-1} line probably contains intensity from phenylalanine (Phe), tyrosine (Tyr), and tryptophan (Trp), although the line at 1552 cm^{-1} is likely from tryptophan alone and may be used to gauge the intensities of other lines from tryptophan]. These lines are not particularly sensitive to the environment of the aromatic amino acid side chains or to the conformation of the polypeptide chain.

The strong line near 1450 cm^{-1} arises from deformation vibrations of methylene (CH_2) groups. The region from 1450 down to 1300 cm^{-1} contains many overlapping lines primarily resulting from C-H deformation modes of methyl (CH_3) and methine (CH) groupings. Recently, some understanding of the origins and intensities of these lines has been developed. The lines as a group may be exploited successfully as an internal intensity standard [14].

The tryptophan line near 1360 cm^{-1} is an indicator of the environment of these residues (see Table 4-2). The lack of a sharp, strong peak near 1360 cm^{-1} in curves A or B shows that, for coat-protein molecules in the capsid structure, both tryptophan residues are exposed to solvent and are not surrounded by hydrophobic groups. The lines at 760 and 875 cm^{-1} also from tryptophan, provide corroboration for this conclusion.

Recent correlations developed for filamentous virus coat proteins [14] also permit the numerous weak bands in the 1200-900 cm^{-1} region to be assigned with confidence to aliphatic side-chain vibrations, especially C-C stretching modes. Also, in the same interval is a strong line near 1004 cm^{-1}, which identifies the phenylalanine side chain. Earlier studies suggested little change in the intensity of this line with protein structure; more recently, it has been recognized that the intensity of the 1004 cm^{-1} peak can vary greatly among proteins containing the same numbers of phenylalanine residues.

The intensity ratio of the aforementioned tyrosine doublet at 850 and 830 cm^{-1} (Section I) is diagnostic of the hydrogen bonding environment of the phenolic OH group [10]. The spectrum of MS2 exhibits an intensity ratio

I_{850}/I_{830} that is less than 0.3, indicating that all four tyrosine residues per coat-protein molecule form strong hydrogen bonds with highly negative acceptors, such as the $-CO_2^-$ groups of glutamic (Glu) and aspartic (Asp) acid residues. There are nine such carboxylate groups per coat-protein molecule. It is unlikely that any of the tyrosine residues hydrogen bond to water (for which $I_{850}/I_{830} = 1.3$) or to highly positive groups, such as lysyl $-NH_3^+$ (for which $I_{850}/I_{830} = 2.5$). This internal H bonding could be a significant source of stabilization energy for the capsid structure.

Spectrum of RNA. The Raman spectrum of protein-free RNA from MS2 is shown at several temperatures for H_2O solutions (curves A) and D_2O solutions (curves B) in Figure 4-6.

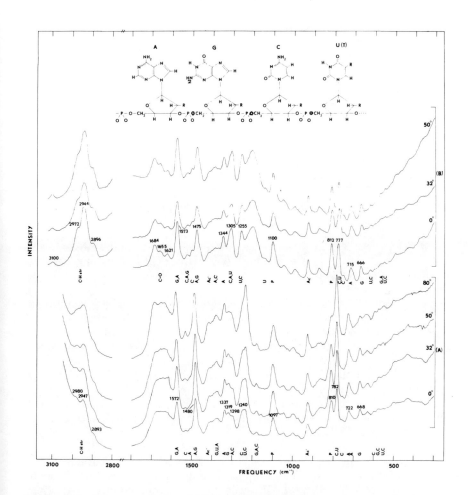

Figure 4-6. Raman spectra of MS2 RNA in H_2O (A) and D_2O (B) solutions. Reprinted with permission from J. Mol. Biol. [24].

A contour of strong scattering resulting from C-H stretching vibrations is seen near 2900 cm^{-1} as was seen for MS2 protein. Between 1600 and 1700 cm^{-1}, scattering from vibrational modes of the RNA bases (involving C=O, C=N, and C=C groups) produces a number of overlapped lines that contain information about the extent of base pairing. This region is best observed in curve B because no solvent interference is encountered. It is important to note that the weak and broad scattering envelope from RNA is unlikely to perturb the position of the amide I line in a superposition of the RNA and protein spectra (compare Figures 4-5 and 4-6).

Between 1600 and 1200 cm^{-1} a number of strong lines are observed resulting from in-plane vibrations of the bases. Among the strongest of these, with their assignments, are 1575 (G, A), 1485 (A, G), 1340 (A), 1320 (G), 1248 (C), and 1238 (U, C). Note that the last two lines will interfere with the interpretation of the amide III region for protein in the spectrum of the virus. The line at 1575 cm^{-1} is the most useful intensity standard in spectra of all nucleic acids [21,22].

The strong line at 1100 cm^{-1} is assigned to the sugar-phosphate linkages of RNA and arises from the symmetric stretching vibration of the PO_2^- portion of the phosphate group. This band is often used as an intensity standard for RNA.

Beyond several weak lines is the important line at 815 cm^{-1}, which is from the C-O-P-C linkage and denotes the conformation of the ribose phosphate chain. The intensity of this line, as measured by I_{815}/I_{1100}, shows that 85% of the nucleotide units of purified MS2 RNA are in the A conformation [11].

Other structurally informative lines in the spectrum of MS2 RNA are near 785 (C, U), 727 (A), and 672 cm^{-1} (G). Many of the remaining lines also have been assigned [24].

The spectra of Figure 4-6 also show the hypochromic and hyperchromic effects mentioned in Section II. As the temperature of the solution of MS2 RNA is increased, the intensities of the hypochromic lines increase. This results from unstacking and unpairing of the RNA bases with temperature (melting). The hypochromic effect is most easily seen for the lines near 1531, 1340, 1240, 785, and 727 in the spectra of H_2O solutions. The weak line at 672 cm^{-1} from guanine residues decreases in intensity with increasing temperature and is therefore hyperchromic. Recently it has been shown that this line is diagnostic of the C3'-endo sugar pucker and anti orientation of the guanosine residue [20,22].

Of particular interest is the line at 815 cm^{-1} which, as mentioned above, measures the fraction of sugar-phosphate groups in the A structure. As temperature is increased (Figure 4-6), the intensity of this line decreases, which shows the melting of the ordered RNA backbone.

The normalized intensities of temperature-sensitive Raman lines may be plotted against temperature to reveal the melting behavior of the various groups. As shown in Figure 4-7, left side, a number of RNA subgroups melt over a narrow temperature range. Included in this category are the RNA purines and pyrimidines, which evidently make a cooperative transition from ordered to disordered states.

A second set of lines shows a more gradual transition versus temperature (Figure 4-7, right side). It is clear that the phosphodiester groups continuously and gradually become disordered and this transition lacks the cooperativity shown above for the bases. A gradual change also is observed for the elimination of base pairing, as revealed by the line at 1657 cm^{-1}. The interpretation of the gradual transitions of the lines at 1475 (G, A) and 1238 cm^{-1} (C, U) is less clear, but these intensities probably depend at least partly on hydrogen bonding, as is the case for the line at 1657 cm^{-1}.

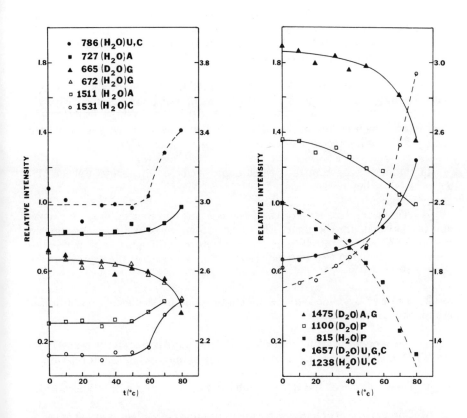

Figure 4-7 Melting profiles of Raman lines of MS2 RNA. Reproduced with permission from J. Mol. Biol. [24].

Spectrum of MS2 phage. The spectra of MS2 in H_2O (curve A) and D_2O (curve B) are shown in Figure 4-8. The spectra are the same for solutions containing 0.75 or 0.1 M KCl and are, within experimental uncertainty, equal to a summation (70% protein and 30% RNA) of the spectra of the capsid and pure RNA. This shows that the inclusion of the RNA in the capsid makes no detectable change in the secondary structures or interactions of either component. In particular, no lines are observed at 1360 or 855 cm^{-1}, which shows that the environments of tryptophan and tyrosine are the same as in the RNA-free capsid. Similar conclusions for the secondary structures of the protein and RNA are drawn from the amide I line near 1660 and the phosphodiester line near 815 cm^{-1}.

The spectrum of the phage in 0.75 M KCl solution shows no change as temperature is increased to 50°C. It can be concluded that the RNA structure in the phage is stabilized in comparison with the changes seen in Figure 4-7B for protein-free RNA. Above 55°C, the protein separates from the RNA and precipitates, leaving a spectrum of pure RNA.

For MS2 in 0.1 M KCl solution some significant changes take place between 40 and 50°C as revealed by the RNA lines at 814, 1324, and 1335 cm^{-1}. This indicates that the lower ionic strength destablizes some portion of the RNA structure in the phage particles.

The Raman spectrum of heat denatured (HD) particles, produced by heating the phage at 50°C in 0.1 M KCl for 10 min, was identical with the spectrum of the native phage. It appears that the partial release of RNA from the capsid, which occurs during HD formation, does not change significantly the structure or interactions of either the protein or the RNA.

In summary, the Raman spectroscopic study of the bacteriophage MS2 gave the following information:

1. The secondary structures of the coat-protein molecules in the capsid and in the virus are the same. Each coat-protein molecule contains approximately 60% β-sheet and 40% irregular structure, although a small percentage of α-helix (ca. 10%) also may be present. The structure of the capsid is stable up to 50°C.

2. The two cysteine residues per coat-protein molecule are present as S-H groups that are exposed to the solvent.

3. The secondary structure of MS2 RNA is approximately the same in the virus, HD particle, and protein-free state and contains 85% A conformation.

4. The structure of the RNA is stabilized by inclusion within the capsid.

5. The two tryptophan residues per coat-protein molecule are exposed to the solvent.

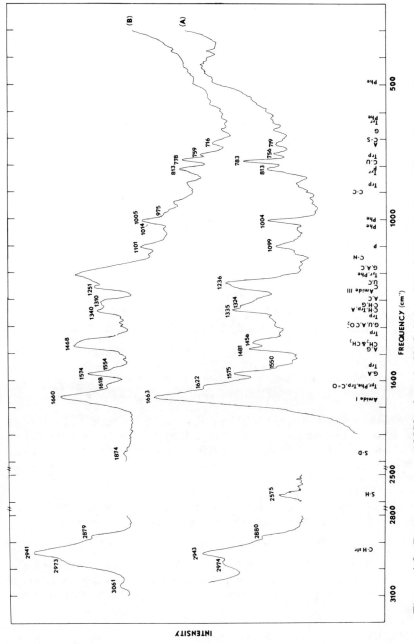

Figure 4-8. Raman spectra of MS2 phage in H_2O (A) and D_2O (B) solutions. Reproduced with permission from J. Mol. Biol. [24].

6. The four tyrosine residues per coat-protein molecule form strong hydrogen bonds with negative acceptor groups (most likely $- CO_2^-$ from Glu or Asp).

B. Turnip Yellow Mosaic Virus

This discussion now turns to a plant virus that is remarkably similar to MS2 in gross structure [1,2]. Turnip yellow mosaic virus (TYMV) is also an isometric virus composed of 180 coat-protein molecules (mol wt 20,113) and one molecule of single-stranded RNA (mol wt 1.91×10^6). However, this RNA is degraded easily while in the virus particle. The TYMV is composed of 65% protein and 35% RNA by weight; it is more massive than the MS2 particle. Besides differences in amino acid composition (TYMV has fewer aromatics), TYMV RNA is especially rich in cytosine as compared with MS2 RNA (38.3% vs. 25.5%).

The RNA-free capsid and protein-free RNA also may be prepared from TYMV so the study of TYMV by Raman spectroscopy has followed the same logical development as used for MS2. Therefore it is not necessary to present a line by line analysis as given for MS2 and this discussion can move directly to the important conclusions as drawn from the intensities and frequencies of the lines. The results and conclusions that follow are taken from ref. 25 and 26, which contain supporting arguments and detailed references to the literature.

The positions of the amide I and III lines in the capsid spectrum and the amide I line in the virus spectrum suggest that the coat-protein molecule in both states contains nearly equal fractions of β-sheet and irregular structures (ca. 45%), with at most a small fraction (ca. 10%) of α-helix.

From the relative intensity of the sugar-phosphate line at 815 cm^{-1}, it can be deduced that the RNA in the virus particle is about 77% in the A-helical form and this is not changed by the introduction of up to 200 chain scissions. The secondary structures of both RNA and protein are stable in the virus particle at temperatures up to 54°C.

The four cysteine residues per coat-protein molecule are in the S-H form (no S-S group is detected) and these readily exchange with solvent D_2O.

The three tyrosine residues per coat-protein molecule are distributed in one of two ways. Either all are hydrogen bonded with solvent water, or one tyrosine forms a strong hydrogen bond to a $-CO_2^-$, a second forms a moderate bond to H_2O, and the third accepts a strong hydrogen bond from $-NH_3^+$.

Both tryptophan residues per coat-protein molecule are exposed to solvent water.

One important property of Raman spectroscopy not discussed for MS2 is the ability to detect protonation of molecular subgroups, such as cytosine and adenine. If one of these basic groups is located near a carboxylate ($-CO_2^-$) group from Asp or Glu, the normal pK_a value (4.2 and 3.7, for cytosine and adenine respectively) would be significantly increased and protonation might occur near neutrality. This mechanism has been proposed by Kaper as a source of stability for the virus particle [1]. This possibility was explored by recording the Raman spectrum of TYMV as a function of pH and observing the diagnostic lines from protonated and unprotonated cytosine and adenine. Nonprotonated (neutral) cytosine has strong lines at 1245 and 1295 cm^{-1}, whereas protonated cytosine has a very strong line at 1255 cm^{-1}. Although uracil residues also scatter in this region, protonation produces a definite reduction of intensity near 1295 cm^{-1} and an increase near 1255 cm^{-1}. These changes are clearly seen for pure ribosomal RNA (from E. coli) and to a lesser degree for TYMV as pH is lowered [26]. Because the protein in TYMV also scatters in these regions, the changes are less dramatic for TYMV than for pure ribosomal RNA. Nevertheless, it may be concluded from the data that by pH 4.8, a significant fraction of the cytosine residues are protonated and therefore that the pK_a for cytosine has been increased within the virus.

The most useful lines for the estimation of neutral adenine are those at 1480 and 1575 cm^{-1}, which are strong for unprotonated adenine. The spectrum of TYMV shows decreases in the intensities of these lines as pH is lowered [26], which suggests that some adenine residues protonate at pH 4.8 and that a significant fraction protonate at pH 4.1. As in the case of cytosine residues, the pK_a values for adenine residues in the TYMV particle have been increased by interactions with amino acid residues.

C. Tobacco Mosaic Virus

The tobacco mosaic virus (TMV) particle differs in many important respects from MS2 and TYMV. The TMV is rod shaped and is composed of over 2000 identical coat-protein molecules (mol wt 17,420) that form a helical array enclosing one molecule of single-stranded RNA (mol wt 2.2 x 10^6). Only 5% of the virus particle is RNA [1-3]. More is known about all levels of the structure and assembly of TMV than for other viruses, and a full description is beyond the scope of this chapter. Therefore this discussion is restricted to those points germane to the information obtainable from Raman spectra.

As in the case of MS2 and TYMV, protein-free capsids can be isolated for TMV. In addition, a precursor of the capsid, called a disk (which contains 34 coat-protein molecules), is also available [3]. Therefore, we recorded Raman spectra of TMV particles, capsids, disk, and RNA (Figure 4-9). The major results and conclusions follow [27].

The secondary structure of the coat-protein molecule is the same for the disk, the capsid, and the mature virus particle and is 45 ± 5% α helix, 45 ± 5%

irregular, and 0-20% β sheet.

Two of the three tryptophan residues per coat-protein molecule in the capsid, disk, and virus reside in hydrophobic regions, whereas the third is in contact with solvent water.

There are many possible distributions of the four tyrosine residues per coat-protein molecule among the three distinguishable hydrogen-bonding states mentioned above. The observed tyrosine doublet intensity ratio excludes several of these from consideration and those that remain consistent with the

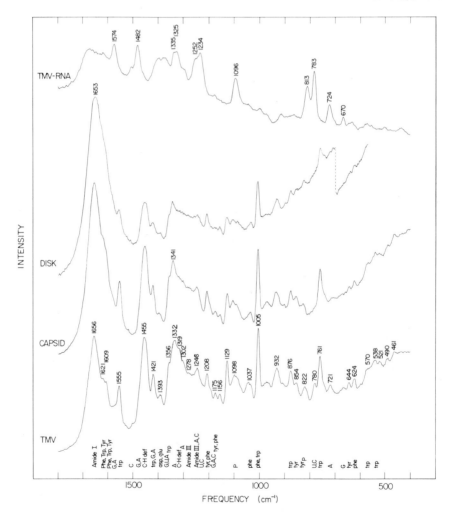

Figure 4-9. Raman spectra of TMV, capside, disk, and RNA. Reprinted with permission from Biochemistry 20, 7449-7457, copyright (1982) American Chemical Society.

Raman data have been discussed [27]. Although the capsid and the virus share the same distributions, those available to the disk are quite different.

This appears to be the first instance in which a feature of the structure of a coat-protein molecule differs depending on aggregation state and this result is not expected from X-ray diffraction data. The answer to this anomaly probably lies in the method of preparing the disk, which employs 0.8 M KCl at pH 8. It is possible that Cl$^-$ may approach tyrosine OH groups and replace water molecules. Such a change in the hydrogen bonding of the OH group could account for the observed results. In comparison with structures derived from X-ray diffraction data, it appears that the distribution in which all tyrosine OH groups hydrogen bond with solvent water is the most likely for coat-protein molecules in all three aggregation states [27].

The spectrum of protein-free TMV RNA is similar to those of other RNAs. The intensity of the 815 cm^{-1} line predicts that about 80% of the sugar-phosphate groups are in the A form. The relatively low mole percentage of cytosine residues (20.0%) is reflected in the spectrum by the relative weakness of the lines from cytosine near 1295 and 1245 cm^{-1} when compared with the corresponding lines in the spectra of E. coli rRNA (21.5%), MS2 RNA (25.5%), and TYMV RNA (38.3%C) (see Section IV).

The analysis of the RNA lines in the Raman spectrum of TMV is difficult because of the small fraction of RNA in the virus particle (1/20 by weight). However, the strongest lines from the RNA bases, including some of those that normally show Raman hypochromism, are distinguished in the spectrum of TMV. For example, lines at 721, 789, 1248, 1302, and 1531 cm^{-1} display more intensity than expected. This is consistent with the absence of hypochromism for the extended RNA molecule of the virion. It appears that the RNA in the virus is configured so as to prevent the stacking of bases that normally generates Raman hypochromism in single-stranded RNA.

A second difference in structure between free and encapsidated RNA is revealed by the high-resolution spectrum in the 800-840 cm^{-1} region, which shows, besides one tyrosine line, two lines from the C-O-P-O-C lingkage [27]. These lines, at 820 and 812 cm^{-1}, show that two distinct configurations exist for the three nucleotide residues associated with each coat-protein molecule. Probably two nucleotide residues exist in one configuration and the third exists in the alternative configuration. All of these conclusions confirm and extend the interpretation from X-ray diffraction data.

D. Bacteriophage P22

The bacteriophage P22, which infects Salmonella typhimurium, has an isometric capsid like MS2 and TYMV, but this is attached to a tail structure that functions during the infection process. Approximately 420 coat-protein molecules (mol wt 55,000) make up the capsid. Several other proteins also exist in the mature phage along with one molecule of double stranded DNA (mol wt 2.7 x 10^7). Therefore the phage particle is about 54% DNA and 46% protein and is much larger and more massive than MS2.

The assembly of P22 is also different from assemblies of MS2 and TYMV and involves a scaffolding protein that interacts with and orients the coat-protein molecules. This process produces a structure called a procapsid, which then absorbs the DNA molecule. During DNA packaging, the scaffolding protein is expelled and recycled. Finally, the tail structure is built up and the mature phage particle is complete.

The complexity of P22 morphogenesis is reflected in the number of structures available for Raman study (Figure 4-4) and by the number of questions to be answered about the interactions involved in assembly. Some of the salient features of the component structures deduced from Raman spectra were reviewed in Section II, E. Further details are given in ref. [23].

E. Filamentous Bacteriophages

Pf1 and fd. The filamentous viruses Pf1 and fd were the first DNA viruses to be studied by Raman spectroscopy [28]. Both have diameters of 6 nm, but they differ in length (1950 nm for Pf1 and 900 nm for fd) and helical symmetry. Both are composed of many relatively small coat-protein molecules (mol wt 5000) and one molecule of DNA, which accounts for a maximum of 12% by weight of the virus particle. Each filamentous virus also contains a few molecules of a minor protein that functions in adsorption of the virus to the host.

Raman spectra indicate that coat-protein molecules in both Pf1 and fd have highly α-helical structures [14,28], although the Pf1 subunit is more nearly a pure α-helix than the fd subunit. As the temperature is increased from 30 to 75 °C, some changes in the hydrogen bonding of the α-helix are observed but no major change in structure takes place, provided prolonged heating is avoided. Lowering the temperature removes these perturbations but enables in Pf1 a structure transition of the whole filament to an altered helical state [14].

The strong lines seen for tryptophan and phenylalanine residues in the spectrum of fd (757, 878, 1005, 1012, 1362, and 1560 cm^{-1}) are missing from the spectrum of Pf1, which confirms the differences in amino acid composition and clearly distinguishes these phage spectroscopically (see Section IV).

Despite the low fractions of DNA in Pf1 and fd, Raman lines from DNA can be discerned near 1580 and 785 cm^{-1}. These and other lines of DNA also can be enhanced by computer averaging of the spectra. Most recent results reveal that classical A and B backbone conformations are absent from encapsidated DNA in all of the filamentous viruses studied, including If1 and IKe (which contain the same helical symmetry as fd, ie, class I symmetry) and Xf and Pf3 (which contain the same helical symmetry as Pf1, ie, class II [14]).

Other unusual features revealed in the Raman spectra of Pf1 and fd involve the amino acid side chains. For example, all tyrosines of Pf1 and fd are involved in strong hydrogen bonding to positive donor groups and are not accessible to solvent. Similarly, the tryptophan residues of fd, If1, and IKe are sequestered in hydrophobic pockets, but in the class II structures Xf and Pf3, tryptophans are in unusual environments, very likely juxtaposed with DNA bases and possibly involved with the bases in aromatic ring stacking interaction.

<u>Detailed structure comparisons among six filamentous viruses.</u> A comparative and detailed Raman study has been carried out on six filamentous viruses, three of which exhibit class I symmetry (fd, If1, and IKe) and three of which are in class II (Pf1, Xf, and Pf3) [14]. Spectra typical of the six viruses are shown in Figure 4-10, which demonstrates clearly the uniqueness of the Raman spectrum of each virus, although all share the same morphology and many structural similarities. The Raman spectra of the six viruses have been examined carefully as a function of both solution temperature and ionic strength [14,29]. The important conclusions from this comparative analysis are as follows:

1. The observed order of decreasing α-helicity of coat protein molecules in the native viruses is: Pf1 > IKe > fd $\sim$ If1 > Pf3 > Xf. Here, the Pf1 subunit is 100% helix, and that of Xf is 50% helix. The nonhelical structure in Xf appears to be mostly irregular; ie, virtually no β-sheet structure occurs in the native Xf subunit. The nonhelical structure in Pf3, in contrast, is all β-sheet and accounts for 25% of the protein chain. Nonhelical regions in fd and If1 ($\leq$ 20%), as well as in IKe ($\leq$ 10%) are difficult to categorize specifically but may be β sheet or irregular or a combination of both.

2. Helices of the class II viruses Pf1, Xf, and Pf3 (but not those of the class I viruses fd, If1, and IKe) undergo reversible thermal transitions to β sheets. The transitions involve most of the protein structure and are facilitated by prolonged heating above 60°C or by reduction of the solution ionic strength. Examples of these $\alpha \rightarrow \beta$ transitions are shown in Figure 4-11 where the Raman difference spectrum between two temperatures is displayed for Pf3 and Xf. In each difference spectrum, the strong positive band near 1645 cm^{-1} (amide I, α helix) and the strong negative bands near 1670 cm^{-1} (amide I, β sheet) and 1234 cm^{-1} (amide III, β sheet) reveal the magnitude of the secondary structure transition.

3. A structure transition in Pf1, previously observed by X-ray diffraction and calorimetric methods and involving a change in filament helical symmetry, is also observed by Raman spectroscopy of Pf1. Similar transitions are also revealed in class I viruses. The transitions are ionic strength dependent and involve changes in the environments of aliphatic and aromatic amino acid side chains of coat-protein molecules.

4. All virus structures except Pf3 are sensitive to salt concentration in the range 0.01-0.50 M NaCl. Generally, the unstacking of aromatic amino acid

residues (phenylalanine and tryptophan) and the introduction of gauche conformers of aliphatic chains (lysine, aspartic acid, etc) are promoted by lowering the salt concentration.

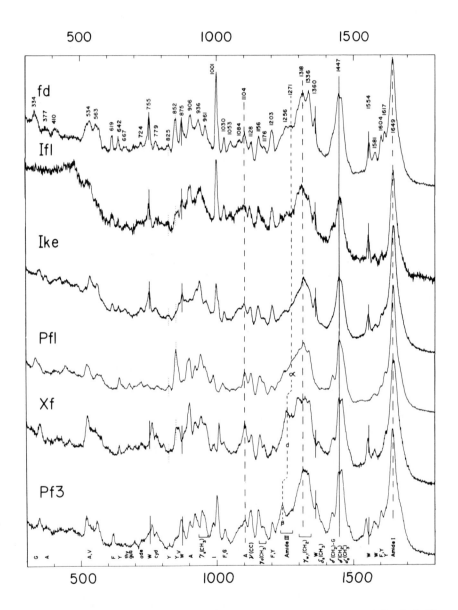

Figure 4-10. Raman spectra of the filamentous viruses fd, If1, IKe, Pf1, Xf,
and pf3. Reproduced with permission from J. Mol. Biol. [14].

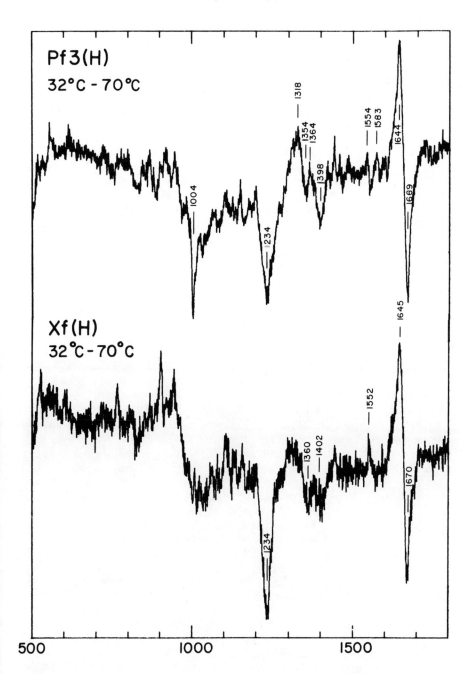

Figure 4-11. Raman difference spectra of Pf3 virus and Xf virus between 32 and 70°C. Reproduced with permission from J. Mol. Biol. [14].

5. The sugar-phosphate backbones encountered for aqueous or fibrous DNA (A, B, and Z helices) do not occur in any of the filamentous viral DNAs, which exhibit unusual DNA conformations. The bases in class I virions are deduced to be unstacked and unpaired, more similar to those occurring in heat-denatured DNA than to classical DNA secondary structures.   Base configurations in Pf1 and Xf are similar to one another, different from class I structures, and unlike any known DNA conformation.   Syn conformers of guanine may be present in the DNA of Pf3.

6. The two phenylalanine residues near the carboxyl terminus in the subunit of each class I virus are deduced to be stacked with one another and/or with DNA bases. Such stacking interactions may stabilize subunit-DNA association. The Raman spectra suggest that similar interactions are likely between tyrosines and tryptophans of class II subunits and their corresponding DNAs.

IV. IDENTIFICATION OF VIRUSES BY RAMAN SPECTROSCOPY

The identification of different species of virus by spectroscopic methods has received little attention in the scientific literature. However, a sufficient number of species has now been investigated by Raman spectroscopy to allow this subject to be addressed here.

Infrared spectroscopy is well known to provide a fingerprint of a pure molecular compound. The complicated nature of the vibrational modes arising from the coupled vibrations of the molecular skeleton produces a unique set of infrared absorption bands below 1500 cm^{-1}, which is in fact called the fingerprint region of the infrared spectrum. Because Raman spectra have in the recent past been more difficult to obtain than IR spectra they generally have not been exploited for fingerprinting, but this capability (based on the same theory) is fully shared by both infrared and Raman spectroscopy.

For many biological materials, including viruses, the advantages of obtaining a full spectrum (4000-200 cm^{-1}) in aqueous solution, coupled with the strong and unique nature of the lines from the aromatic ring modes of proteins, nucleic acids, and other components, advance Raman spectroscopy as a potentially useful fingerprinting method. Heretofore, this potential has not been exploited. One reason for this is that functionally different viruses may be composed of very similar nucleic acid and protein components, requiring differentiation by minor spectral differences. Another reason is that relatively large quantities of virus are needed for Raman spectroscopy as compared with biochemical or immunological assaying methods.

In this section the spectra of the RNA and DNA-containing viruses presented above are compared and differences arising from composition and structure are discussed. It will be seen that for the examples in hand, different species of virus have significantly different spectra. Only in the situation that a minor mutation separates the variants are the spectra too similar to be easily distinquished.

A. RNA Viruses

The bacteriophage MS2 and the plant virus TYMV have similar ratios of RNA to protein with some clear differences in amino acid and nucleotide composition. Despite these similarities, significant differences are seen in Raman spectra of MS2 and TYMV. A close comparison of Figure 4-6 and figures in ref. 26 will show many qualitative differences, a few of which are as follows. (1) The ratio of the intensities at 1480 (A, G) and 1455 cm^{-1} (C-H deformations) is larger in MS2 than in TYMV. (2) The same is true for I_{1340}/I_{1295}. (3) Major differences in the tyrosine lines near 850 and 830 are observed. (4) The tryptophan lines near 760 and 1555 cm^{-1} are also more prominent in MS2. (5) The intensity at 1248 cm^{-1} (from cytosine) is much greater in TYMV than in MS2, which reflects the known base compositions of these RNAs (see Section III).

It is clear that MS2 and TYMV are easily distinguishable by their Raman spectra.

Tobacco mosaic virus contains a much higher fraction of protein than does MS2 or TYMV, and this is shown in the Raman spectra. The RNA lines near 1575, 1480, 1295, 1240, 1100, and 780 cm^{-1} are missing or greatly reduced in the spectrum of TMV when compared with MS2 or TMYV (see Figure 4-7). Also, the spectrum of TMV shows comparatively strong lines for phenylalanine (1005 cm^{-1}) and tyrosine (1555 and 760 cm^{-1}). Many additional differences could be mentioned (eg, compare lines near 1129 cm^{-1}), but it is clear that the spectrum of TMV is distinctly different from those of MS2 and TYMV.

A counterexample is the bacteriophage R17, which gives a nearly identical spectrum to that of MS2. However, these phage are very closely related and may differ only by a single amino acid residue in the coat-protein molecule.

B. DNA Viruses

The Raman spectrum of the DNA containing bacteriophage P22 is very different from the spectra discussed above. Lines from phenylalanine (1005 cm^{-1}) and tryptophan (1555 and 760 cm^{-1}) have comparatively low intensities, as does the line near 1455 cm^{-1} from C-H deformations. Lines from DNA (eg, 1580, 1489, 1094, and 787 cm^{-1}) are quite strong. This reflects the relatively high fraction of nucleic acid in P22 as compared with MS2 and TYMV.

The filamentous bacteriophages fd, If1, IKe, Pf1, Xf, and Pf3 Figure 4-10) all contain about 12% or less DNA but each spectrum is different from the others and from the spectra of P22 and the RNA viruses.

Among the most significant differences between spectra of any two filamentous viruses are Raman lines of the aromatic amino acid residues. For example, fd gives sharp, clear lines from tryptophan at 1560 and 757 cm^{-1}, which are missing in the spectrum of Pf1. The very strong and sharp line at 1005 cm^{-1} seen for fd is missing for Pf1, indicating a major difference in the content of phenylalanine. In fact, Pf1 is known to lack tryptophan and phenylalanine residues, whereas fd does contain these residues. This example also shows that Raman spectra may provide at least a semiquantitative measure of the content of aromatic amino acids in the proteins of viruses. The regions 1200-1050 and 700-500 cm^{-1} also show major differences among all filamentous viruses (Figure 4-10).

The identification and possible classification of viruses by Raman spectroscopy is therefore demonstrated for a number of examples. It should be possible to develop or adapt computer programs (such as are now available for infrared spectra) in which all the lines in the spectra of known viruses are compared with the Raman lines of an unknown virus. The computer could be made to indicate the closest match and other significant matches and could identify strong lines or missing key lines (eg, 1005 cm^{-1} for Xf and Pf1). With new Raman instruments coupled with computers, Raman spectroscopy has the potential for the rapid identification of viruses and possibly also of bacteria and other microorganisms.

V. SOME NEW METHODS IN THE RAMAN SPECTROSCOPY OF VIRUSES

A. Computer-Difference Spectroscopy

Pf1. Many of the structural conclusions reached from Raman spectra of viruses require the careful comparison of two or more spectra. This kind of comparison is demonstrated by the difference spectra of Figure 4-11, where the spectrum of the virus at high temperature is subtracted from that at low temperature to show all of the Raman frequency and intensity changes that occur when the temperature of the virus is changed. The process of spectral subtraction is extremely tedious to carry out manually but is achieved rapidly and accurately by a computer interfaced to the spectrometer. The subject of Raman difference spectroscopy actually entails some elaborate theoretical consideration and its practice can require rather sophisticated experimental protocols. A thorough review of this subject has been given recently by Laane [30]. Here, some of the power of the method is demonstrated for virus applications.

Figure 4-12 shows the Raman spectra of Pf1 virus at three different salt concentrations, designated high (H), medium (M), and low (L). Some of the Raman intensity and frequency changes induced by salt are evident simply from inspection of the three spectra; yet quantitative comparisons are hardly permitted by visual examination. The lowest trace in the figure shows the computed difference spectrum corresponding to the change from high salt (minuend) to low salt (subtrahend). The computed difference spectrum shows clearly the shift of intensity from 1306 to 1341 cm^{-1} and the large intensity decrease in amide I (1651 cm^{-1}) attendant with lowering the salt concentration. Further, the computer permits rapid normalization of the ordinate scale for quantitative intensity comparison. Here the difference spectrum and high-salt spectrum are normalized to the same scale (8166 counts) so that the

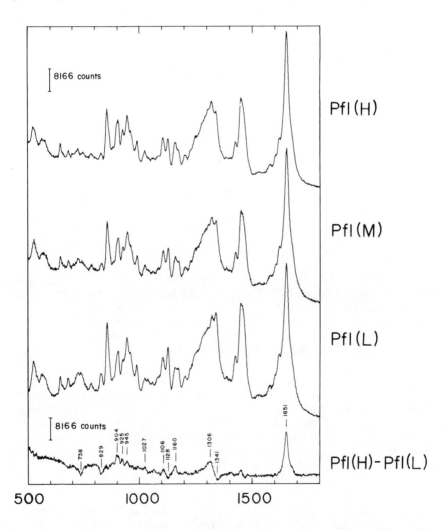

Figure 4-12. Raman spectra of Pf1 in high- (H), medium- (M), and low- (L) salt solutions, and the difference spectrum between H and L. Reproduced with permission from J. Mol. Biol. [14].

former accurately reflects the percentage or fractional change in intensity of the latter.

Reliable structural conclusions can be reached from such data. In this instance, a substantial fraction of the methylene groups of aliphatic amino acid side chains are converted from trans (1306 cm^{-1} line) to gauche conformers (1341 cm^{-1}), and α-helix amide I intensity is reduced by 20%, attributed to unbinding of Na$^+$ groups from peptidyl C=O sites [14].

CCMV. As a second example, consider the cowpea chlorotic mottle virus (CCMV), an RNA plant virus in the bromovirus group [1,2]. Specific RNA and protein interactions are suspected in the virion and, if present, should be evident from comparison of Raman spectra of the virus with its isolated capsid and RNA components. However, because only a small percentage of protein and nucleic acid groups is expected to engage in such nucleoprotein interactions, sensitive detection of spectral band shifts or differences is required.

Figure 4-13 compares the Raman spectrum of the capsid (top) with the difference of CCMV and its RNA (middle). Although the two spectra are very similar, computer subtraction of the middle spectrum from the top spectrum (ie, the "second difference") reveals subtle but clear-cut intensity and frequency shifts that can be interpreted as evidence for RNA-protein interactions in the virion [31].

Finally, Figure 4-14 shows the power of computer difference spectroscopy to reveal three band intensity changes (at 1225, 1297, and 1656 cm^{-1}) accompanying the pH-induced expansion of the CCMV particle. Comparison with pH effects on RNA shows that these band intensity changes are the result primarily of deprotonation of cytosine and adenine rings of viral RNA with swelling [31].

B. Raman Band Deconvolution

Curve fitting techniques, factor analysis, and difference spectroscopy are among the most commonly used methods for analysis of multicomponent bands in vibrational spectra [31]. More recently, deconvolution techniques have provided a convenient and powerful alternative to improving experimental resolution of intrinsically overlapped bands in Raman spectra [32,33]. Here, we illustrate the use of a constrained, iterative deconvolution procedure for the quantitative analysis of certain nucleic acid and protein Raman bands in spectra of Pf1 and Pf3 viruses.

Deconvolution of the amide I band of filamentous virus Pf1. Figure 4-15 compares the observed and deconvoluted Raman bands of Pf1 in the region of the intense amide I protein mode, which dominates the 1500-1800 cm^{-1} interval of the filamentous virus spectrum (see Section III, E). The spectrum shown is unsmoothed, but the signal to noise ratio has been enhanced by signal averaging, and the underlying spectrum of liquid water has been subtracted by computer-difference spectroscopy [14].

Deconvolution clearly identifies the very intense 1648 cm^{-1} amide I band of the protein α-helix conformation and reveals as well the absence of amide I bands between 1655 and 1675 cm^{-1} that may be assigned to either β-strand or irregular conformations. The deconvolution confirms the uniformity of α-helical secondary structure of the coat protein in native Pf1 virus. The weak band resolved at 1677 cm^{-1} is assigned to the carbonyl group stretching modes of the bases of viral DNA and to the glutamine side chains of viral coat

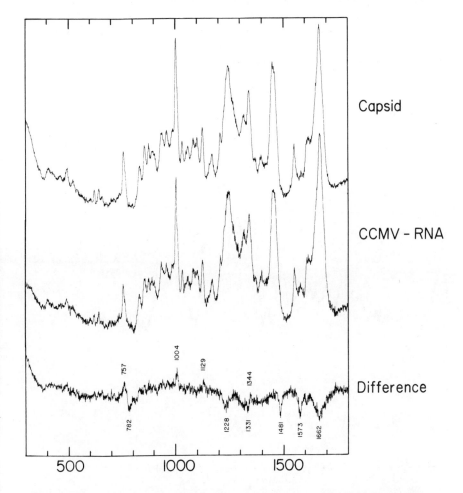

Figure 4-13. Raman spectra of CCMV capsid (top), of the difference between CCMV virus and CCMV RNA (middle), and the second difference between top and middle spectra (bottom). Reprinted with permission from Biochemistry 23, 4301-4306, copyright (1984) American Chemical Society.

protein. The still weaker features at 1570, and at 1596 and 1622 cm^{-1}, are assigned, respectively, to aromatic ring modes of the DNA purines and to tyrosines of the coat protein [33].

Deconvolution of the complex amide I band of the filamentous virus Pf3. The filamentous bacterial virus Pf3 undergoes a structure transition in which the coat-protein conformation is largely converted from α helix to β strand when the temperature is increased to 60°C [29]. Quantitative estimates of the percentages of α and β structures that coexist at various stages of the transition cannot be made from the unrefined Raman data because of extensive overlap of the respective amide I bands. However, it is possible to obtain quantitative estimates by Fourier deconvolution of the complex band envelope as shown in Figure 4-16. Here amide I components are resolved clearly at

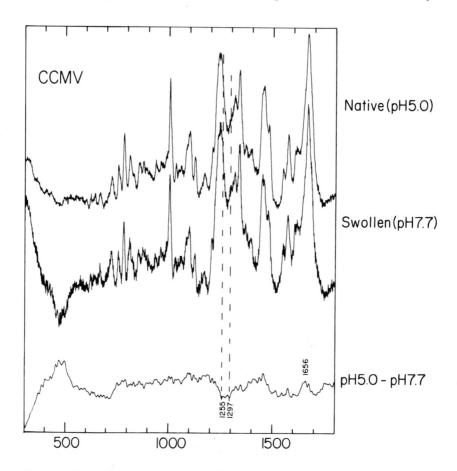

Figure 4-14. Raman spectra of CCMV in native (top) and swollen states (middle) and the difference spectrum. Reprinted with permission from Biochemistry 23, 4301-4306, copyright (1984) American Chemical Society.

1646 cm^{-1} and 1667 cm^{-1} (α-helix and β-strand structures, respectively). We note that irregular protein conformations can be excluded from consideration because of the characteristic amide III Raman scattering of Pf3 [14] not shown in Figure 4-16. The weaker bands in the deconvoluted spectrum are assigned as follows: 1564 (tryptophan), 1582 (DNA purines and phenylalanine), 1609 (phenylalanine), 1619 (tryptophan), and 1690 (glutamine).

The integrated intensities of the two amide I bands indicate a 60:40% ratio of α to β structure in Pf3 at the conditions of Figure 4-16.

The examples given above illustrate the use of constrained, iterative Fourier deconvolution procedure [33] to separate overlapping Raman bands in spectra of native and structurally perturbed viruses for analysis of protein and nucleic acid constituents. This deconvolution method permits several-fold improvement of spectral resolution and allows quantitative treatment of the Raman data. Weak shoulders that cannot be resolved instrumentally from

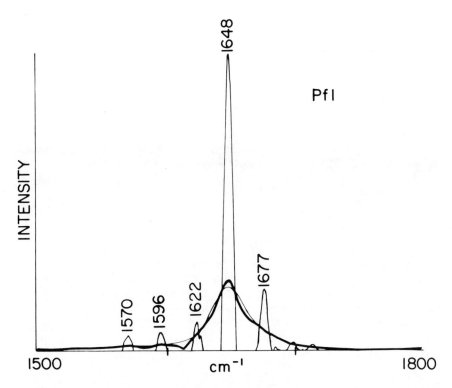

Figure 4-15. Comparison of observed (heavy line) and deconvoluted spectra (light line) of native Pf1 filamentous virus in the region of amide I Raman scattering. Reproduced from the Biophysical Journal (1984) 46, 763-768 by copyright permission of the Biophysical Society.

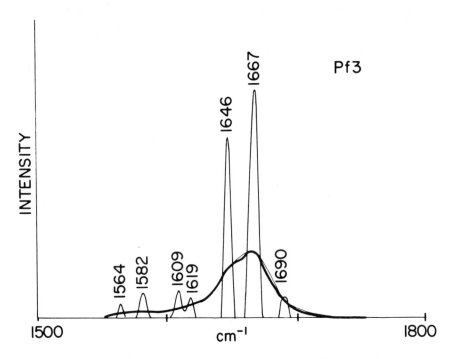

Figure 4-16. Comparison of observed (heavy line) and deconvoluted spectra (light line) of Pf3 filamentous virus in the region of amide I Raman scattering. The native virus structure was perturbed by heating the sample at 60°C for 2 h to promote the α to β structure transition. Reproduced from the Bio-physical Journal (1984) 46, 763-768 by copyright permission of the Biophysical Society.

nearby strong Raman lines are distinguished clearly in the deconvoluted spectra. The same methods are applicable also to the analysis of conformational structure in individual proteins and nucleic acids. In particular, deconvolution provides a convenient alternative to curve fitting of the amide I band for assessment of protein secondary structure [15] and to difference Raman spectroscopy for the quantitation of nucleic acid secondary structures [22].

ACKNOWLEDGEMENT

The support of the National Institutes of Health (Grants AI 11855 and AI 18758) has made possible much of the work described here and is gratefully acknowledged.

VI. REFERENCES

1. Kaper, J. M. "The Chemical Basis of Virus Structure, Dissociation and Reassembly"; North Holland-American Elsevier: New York, 1975.
2. Knight, C. A. "Chemistry of Viruses", second ed.; Springer-Verlag: New York, 1977.
3. Holmes, K. C. Trends Biochem. Sci., 1980, 5, 4.
4. Carey, P. R. "Biological Applications of Raman and Resonance Raman Spectroscopies"; Academic Press: New York, 1982.
5. Tu, A. T. "Raman Spectroscopy in Biology"; John Wiley and Sons: New York, 1982.
6. Thomas, G. J., Jr. In "Physical Techniques in Biological Research"; Oster, G, ed.; Academic Press: New York, 1971; Vol. 1A, pp. 277-346.
7. Hartman, K. A.; Lord, R. C. and Thomas, G. J., Jr. In "Physico-Chemical Properties of Nucleic Acids"; J. Duchesne, ed.; Academic Press: New York, 1973; Vol. 2, pp. 1-89.
8. Thomas, G. J., Jr. Appl. Spectrosc., 1976, 30, 483.
9. Hartman, K. A.; Clayton, N.; Thomas, G. J., Jr. Biochem. Biophys. Res. Commun., 1973, 50, 942.
10. Siamwiza, M. N.; Lord, R. C.; Chen, M. C.; Takamatsu, T.; Harada, I.; Matsuura, H.; Shimanouchi, T. Biochemistry, 1975, 14, 4870.
11. Thomas, G. J., Jr.; Hartman, K. A. Biochim. Biophys. Acta, 1973, 312, 311.
12. Thomas, G. J., Jr. Biochim. Biophys. Acta, 1970, 213, 417.
13. Lord, R. C.; Yu, N. T. J. Mol. Biol., 1970, 50, 509.
14. Thomas, G. J., Jr.; Prescott, B.; Day, L. A. J. Mol. Biol., 1983, 165, 321.
15. Williams, R. W. J. Mol. Biol., 1983, 166, 581.
16. Lord, R. C.; Thomas, G. J., Jr. Spectrochim. Acta, 1967, 23A, 2551.
17. Lafleur, L.; Rice, J.; Thomas, G. J., Jr. Biopolymers, 1972, 11, 2423.
18. Small, E. W.; Peticolas, W. L. Biopolymers, 1971, 10, 69.
19. Erfurth, S. C.; Kiser, E. J.; Peticolas, W. L. Proc. Natl. Acad. Sci. U.S.A., 1972, 69, 938.
20. Thamann, T. J.; Lord, R. C.; Wang, A.H.J.; Rich, A. Nucleic Acids Res., 1981, 9, 5443.
21. Prescott, B.; Steinmetz, W.; Thomas, G. J., Jr. Biopolymers, 1984, 23, 250.
22. Benevides, J.M.; Thomas, G.J., Jr. Nucleic Acids Res., 1983, 11, 5747.
23. Thomas, G. J., Jr.; Li, Y.; Fuller, M. T.; King, J. Biochemistry, 1982, 21, 3866.
24. Thomas, G. J., Jr.; Prescott, B.; McDonald-Ordzie, P. E.; Hartman, K. A. J. Mol. Biol., 1976, 102, 103.
25. Turano, T. A.; Hartman, K. A.; Thomas, G. J., Jr. J. Phys. Chem., 1976, 80, 1157.
26. Hartman, K. A.; McDonald-Ordzie, P. E.; Kaper, J. M.; Prescott, B.; Thomas, G. J., Jr. Biochemistry, 1976, 17, 2118.
27. Fish, S. R.; Hartman, K. A.; Stubbs, G. J.; Thomas, G. J., Jr. Biochemistry, 1981, 20, 7449.
28. Thomas, G. J., Jr.; Murphy, P. Science, 1975, 188, 1205.

29. Thomas, G. J., Jr.; Day, L. A. Proc. Natl. Acad. Sci. U.S.A., 1981, 78, 2962.
30. Laane, J.   In "Vibrational Spectra and Structure"; Durig, J. R., ed.; Elsevier Scientific Publishers: Amsterdam, 1983, Vol. 12, pp. 405-467.
31. Verduin, B. J. M.; Prescott, B.; Thomas, G. J., Jr. Biochemistry, 1984, 23, 4301.
32. Kauppinen, J. K.; Moffatt, D. J.; Mantsch, H. H.; Cameron, D. G. Appl. Spectrosc., 1981, 35, 271.
33. Thomas, G. J., Jr.; Agard, D. A. Biophys. J., 1984, 46, 763.
34. Li, Y.; Thomas, G. J., Jr.; Fuller, M.; King, J. Prog. Clin. Biol. Res., 1981, 64, 271.

5. THE CHEMOTAXONOMIC CHARACTERIZATION OF

MICROORGANISMS BY CAPILLARY GAS CHROMATOGRAPHY

AND GAS CHROMATOGRAPHY–MASS SPECTROMETRY

Alvin Fox and Stephen L. Morgan

I. INSTRUMENTAL METHODS FOR BACTERIAL CHARACTERIZATION

Capillary gas chromatography (GC) or capillary GC coupled to mass spectrometry (GCMS) is a powerful technique for the direct characterization of microorganisms. In comparison to traditional techniques for the analysis of structural components or metabolic products excreted by microorganisms GC offers speed, specificity, and sensitivity. The general application of GC and GCMS for the chemotaxonomic characterization of microorganisms has been the subject of several extensive reviews [1-4]. This chapter is more specific in scope and limited primarily to a discussion on structural differentiation of microorganisms by GC and GCMS, a review of the current literature, and a report on progress from our laboratories.

Analysis of structural components rather than metabolic products does not require the organisms to be viable, and lengthy secondary, and possibly primary, culturing may be avoidable. These aspects may be important in the analysis of fastidious organisms that are difficult or impossible to isolate or that have slow growth characteristics. Table 5-1 lists several such fastidious pathogenic organisms, their associated diseases, and cultivation properties.

The mass spectrometer is a critical component of this chemotaxonomic analytical approach because GC with nonselective detection (for example, flame ionization) does not permit unequivocal identification of chromatographic peaks. Identification of components based only on chromatographic retention time is especially difficult when the number of possible constituents is quite large and where closely related microorganisms can have unique constituents with similar retention times. The coupling of capillary GC to mass spectrometry (MS) improves selectivity and allows the identification or at least partial characterization of the chemical nature of the constituents of interest [11].

The combined technique of GCMS employing high-resolution capillary columns can separate and identify volatile components of complex mixtures rapidly. Unfortunately, macromolecular components of bacterial cells are not volatile. Approaches for volatilizing a biopolymer prior to GCMS include the complementary methods of chemical derivatization or thermal fragmentation

135

Table 5-1. Several Fastidious Pathogens of Man

Bacterium	Disease	Cultivation Properties	Reference
Mycobacterium tuberculosis	Tuberculosis	3-6 weeks in vitro	5
Mycobacterium leprae	Leprosy	No growth in vitro	5
Treponema pallidum	Syphylis	Tissue culture only	5
Treponema pertenue	Yaws	No growth in vitro	6
Legionella pneumophila	Legionnaires' disease and Pontiac fever	Readily grows in vitro only on highly specialized media containing cysteine; first described in 1977	7,8
Tatlockia (Legionella) micdadei	Pittsburgh pneumonia	Grows on same media as L. pneumophila; first described in 1979	9,10

(pyrolysis). Chemical derivatization usually involves several manual sample treatment steps: hydrolysis to release intact monomeric units, followed by reaction with suitable reagents to inhibit hydrogen bonding or ionic interactions. Clean up of the sample prior to GC may be essential in simplifying the resulting chromatogram and making its interpretation easier. Pyrolysis, on the other hand, can be applied directly to the sample without pretreatment. The biopolymer, however, is thermally degraded into fragments that do not necessarily retain the monomeric structure of the bacterial macromolecule. Derivatization and pyrolysis GC or GCMS both produce a characteristic chromatogram; in the case of pyrolysis, the chromatogram of pyrolysis products, or pyrolyzates, is termed a pyrogram. In the total ion abundance mode, a mass spectrometer can scan rapidly across a wide range of mass to charge ratios and produce a complete mass spectrum for an eluting GC peak. In the selected ion monitoring (SIM) mode, only specific ions that are characteristic of the compound (or compounds) of interest are focused on the detector for increased sensitivity. The ability of derivatization or pyrolysis GCMS to provide characteristic patterns and specific analyses for bacteria and bacterial components is based on the fundamental differences in the structure and composition of bacteria.

II. CHEMICAL SIGNATURES FOR BACTERIA

All bacteria are composed of the same basic structural units, including proteins, carbohydrates, lipids, and nucleic acids. There are, however, fundamental differences in the structural composition of different bacterial groups. For example, Gram-positive and Gram-negative bacteria differ in their cell envelope compositions. Although all bacteria possess a cytoplasmic membrane, Gram-negative bacteria have a much thinner peptidoglycan layer and an additional outer membrane. Different macromolecules may be covalently bound to the peptidoglycan. In Gram-negative bacteria, a lipoprotein may be found connected to the peptidoglycan; in Gram-positive bacteria, teichoic acids or polysaccharides usually are found. Other bacterial groups, such as mycoplasmas, chlamydia, and acid-fast and related bacteria, also differ in their respective cell envelope composition. These differences may be exploited to differentiate bacteria by the direct analysis of specific chemical components that signal the presence of certain types of microorganisms. Table 5-2 lists a few of the many chemical markers that are found in bacteria and their sources.

Peptidoglycan is a major component of most bacterial cell walls. Peptidoglycan comprises between 30 and 70% of the dry weight of Gram-positive cell walls but less than 10% of the dry weight of Gram-negative cell walls [12]. Peptidoglycan consists of a glycan backbone that is a repeating polymer of N-acetylglucosamine and N-acetylmuramic acid. Muramic acid is an amino

Table 5-2. Several Chemical Markers for Bacteria

Compounds	Source	Bacterial Group
Muramic acid and D-amino acids	Peptidoglycan	All pathogenic bacteria
Diaminopimelic acid	Peptidoglycan	Certain Gram-positive, Gram-negative, and all acid-fast and related bacteria
Mycolic acid	Cell envelope	Mycobacteria, cornyebacteria, and Nocardia
Heptose, KDO, and HMA	LPS	Gram-negative bacteria
Dipicolinic acid	Endospores	Certain Gram-positive bacteria
Aminodideoxy-hexoses	Cells	Certain Legionella
Rhamnose	Polysaccharide	Group A Streptococcus
Rhamnose	LPS	Certain Gram-negative bacteria

sugar that is unique to bacteria and does not occur elsewhere in nature; thus, muramic acid is a definitive marker for the presence of bacteria. Attached covalently to the lactyl group of the muramic acid moeity in peptidoglycan are tetra- and pentapeptides consisting of repeating L- and D-amino acids crosslinked by peptide bridges. Diaminopimelic acid (DAP) is also a component of these peptides in many Gram-negative bacteria and in certain Gram-positive bacteria. Diaminopimelic acid is also not found elsewhere in nature. Variation in peptide sequences also occur among bacteria. D-amino acids are occasionally found in other bacterial components, such as the teichoic acids (ribitol or glycerol phosphate-based polymers), but are not synthesized by mammalian enzyme systems. A generalized structure for peptidoglycan is shown in Figure 5-1.

Lipopolysaccharide (LPS) is a macromolecule found only in Gram-negative outer membranes and usually contains several unique sugars, including ketodeoxyoctonic acid (KDO) and heptose [13,14]. A LPS molecule consists of three regions: a lipid A region, a core polysaccharide, and an O antigen. The core and lipid A regions are linked covalently by KDO and heptose residues. Lipid A generally consists of a glucosamine disaccharide with long-chain fatty acids in ester and amide linkage. These fatty acids vary in composition among organisms, but often β-hydroxymyristic acid (HMA) is found in lipid A. Usually, KDO and HMA are not found in structures other than LPS and so can be used as markers for Gram-negative organisms. A generalized structure for the lipid A region of LPS is shown in Figure 5-2.

Mycolic acid is a branched, long-chain, hydroxylated fatty acid found as a covalently bound constituent of a number of macromolecules in the cell envelope of acid-fast and related bacteria (corynebacteria, mycobacteria, and nocardia). Mycolic acids are not found in other bacterial groups and furthermore, their chain length differs among corynebacteria, mycobacteria, and nocardia [15].

The routine taxonomic characterization of microorganisms typically requires the employment of biochemical, serological, and antigenic techniques that not only are time consuming but also are usually not automated. Many of these methods require culturing the suspected microorganism in the laboratory for periods ranging from 24 h up to several weeks. If the microorganism grows readily, it is then taken through a battery of standard classification tests for identification. Some of these tests require subjective interpretation and may not provide direct chemical structure information. Alternatively, the above-mentioned chemical markers are just some of the many examples of structural components that may provide differentiation and/or identification of bacterial groups. For reference purposes, Figure 5-3 presents chemical structures for some of these compounds. Whereas one approach to automatically interpreting structural data from bacteria may be to perform simple fingerprinting and matching of patterns to library data bases, a more directed approach is to base successive taxonomic distinctions on the presence or absence of the related chemical marker. In this manner a taxonomic decision-making process could be developed in a tree structure that is

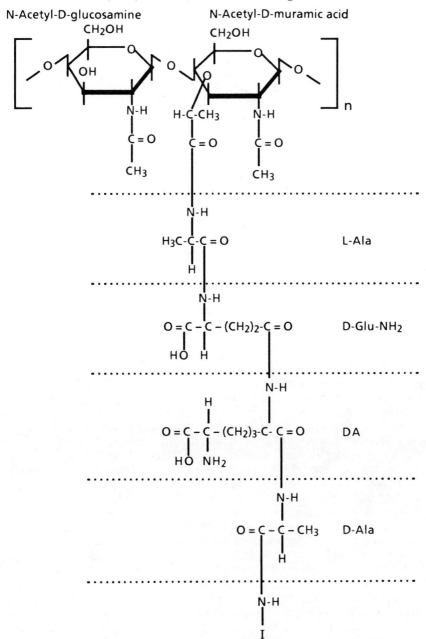

Figure 5-1. Structure of bacterial peptidoglycan. For simplicity, only one linear glycan chain, which would be crosslinked with other chains through peptides bridges, is shown. Abbreviations: L-Ala, L-alanine; D-Glu-NH₂, D-glutamine; DA, diamino acid, here m-diaminopimelic acid; D-Ala, D-alanine; I, interpetide bridge. (Structure adapted from ref. 5.)

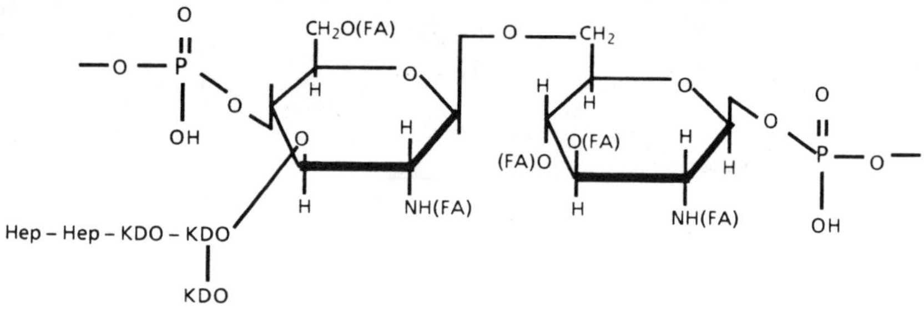

Figure 5-2. Structure of the lipid A region of lipopolysaccharide from Gram-negative bacteria. Abbreviations: FA, fatty acids, including β-hydroxy-myristic acid (HMA); KDO, ketodeoxyoctonic acid; HEP, heptose.

analogous to the more traditional approaches involving physiological and morphological differences.

III. APPLICATION OF DERIVATIZATION GCMS TO DETERMINING GRAM TYPE OF BACTERIA

Traditionally, the first step in differentiating bacteria is, of course, the Gram stain. Any instrumental approach that hopes to complement and possibly supplant these traditional techniques should attempt first to provide information similar to that provided by the Gram stain. The Gram stain divides bacteria into Gram-positive and Gram-negative groups on the basis of the ability of some Gram-positive bacteria to adsorb a purple dye (crystal violet) and, after fixation with iodine solution, to retain it following an extraction with organic solvent. Although the test is dependent on cell envelope structure, it does not always correlate directly with the chemical structure of the microorganism. Indeed, changes in the physical structure of the cell envelope (eg, partial autolysis or a breach in the cell wall) may cause a normally Gram-positive organism to stain Gram negatively despite there being no major changes in the chemical composition of the cell envelope. Presumably, the direct chemical determination of cell envelope composition would not suffer from this drawback. Furthermore, chemical composition information may provide shades of variation ranging from extreme Gram-positive to extreme Gram-negative, thus providing a larger dynamic range for the differentiation of microorganisms. Wiegel [16] has suggested using the term "Gram type" to designate the classification of bacteria based on cell wall structure and type, or based on the presence and absence of characteristic components in contrast to the traditional designation of Gram-positive or Gram-negative based only on tinctorial properties of the cell.

A number of methods have been reported that correlate with the Gram stain. Cerny [17] reported an assay for amino-peptidase, an enzyme found

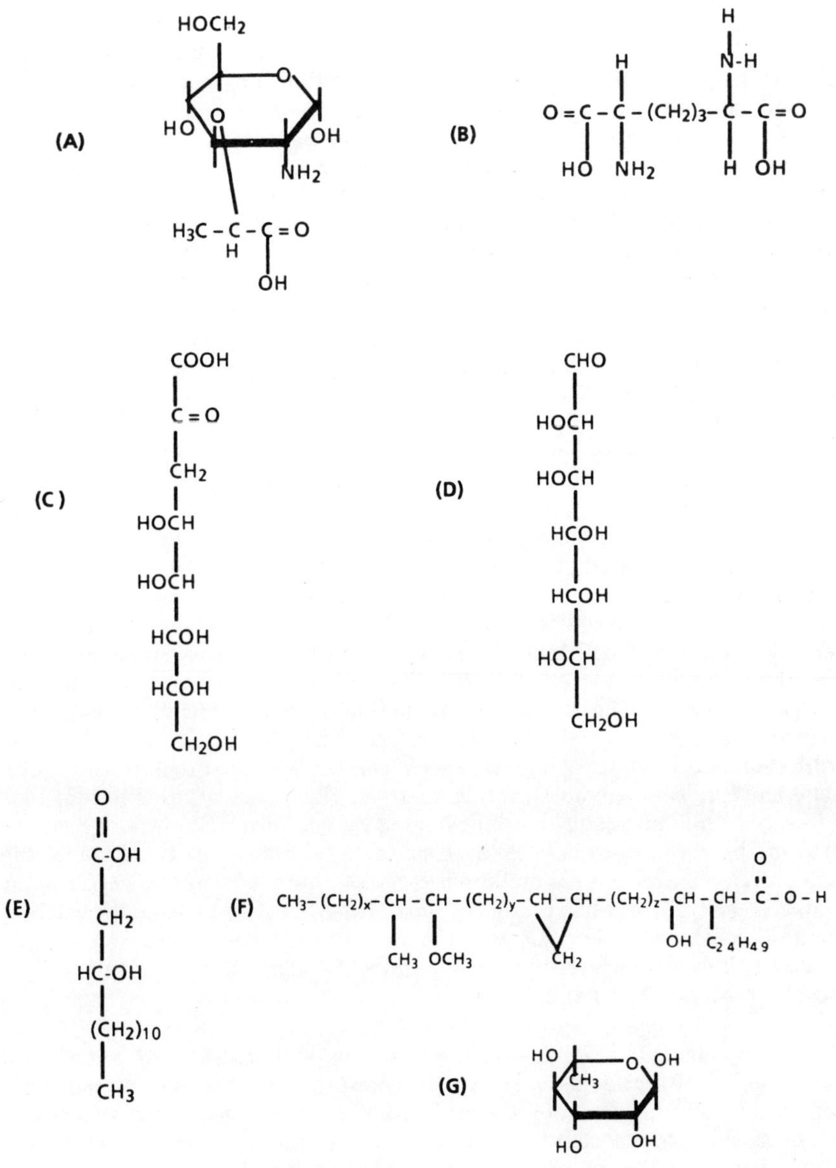

Figure 5-3. Structures of some chemical markers present in microorganisms: (A) muramic acid; (B) m-diaminopimelic acid; (C) ketodeoxyoctonic acid (KDO); (D) heptose; (E) β-hydroxymyristic acid (HMA); (F) mycolic acid; and (G) rhamnose.

primarily in Gram-negative organisms, that distinguishes Gram-negative from Gram-positive bacteria. Wiegel and Quandt [18] reported that the interaction between polymyxin B and outer membranes of Gram-negative bacteria determined by electron microscopy was specific in determining Gram type. Other workers have reported that lysis of the cell wall by potassium hydroxide is an indicator for Gram negativity [19,20]. Carlone et al [21] compared the KOH and amino-peptidase tests with the Gram stain and found good correlation between the three tests. Although tests such as these may be employed in Gram typing, they are indirect methods and do not give precise qualitative and quantitative information about bacterial cell structure. Analytical chemical methods are needed that are rapid, simple to perform, and provide structural information. For example, the presence of lipopolysaccharide (LPS) could be employed as a unique marker for Gram negativity. It may be more valuable, however, not simply to analyze for the presence of LPS based on its biological properties or behavior, but instead to analyze for specific sugars and lipid markers for this biopolymer.

The potential of GC or GCMS to differentiate Gram types of bacteria has not been explored extensively. One obvious possibility may be to employ the relative amount of muramic acid present in bacteria as an indicator of Gram type, because muramic acid is known to be present in higher amounts in Gram-positive compared to Gram-negative bacteria. We have employed two different methods in the GC and GCMS analysis of this monomeric constituent of peptidoglycan, the alditol acetate [22,23] and the aldononitrile acetate methods [24-26]. Both techniques use hydrolysis of the intact cell wall envelope to yield monomeric sugars, followed by derivatization to a suitably volatile compound. The alditol acetate method (Figure 5-4A) involves sodium borohydride reduction, whereas the aldononitrile method (Figure 5-4B) uses hydroxylamine to destroy the anomeric center. Both methods then employ acetic anhydride to produce volatile acetates. The choice of methods depends on the task in hand. The aldononitrile method has been traditionally easier to perform; however, it produces background peaks that may confound the analysis. The alditol acetate method is generally more difficult to perform but produces relatively clean chromatograms. When single components (such as muramic acid) are to be analyzed, the aldononitrile acetate method may be preferred; however, when multiple-component sugars are to be analyzed the alditol acetate method is preferred.

For any derivatization method, an internal standard should be selected that usually is not present in the microorganisms to be analyzed. We generally employ N-methyl glucamine as an internal standard for amino sugars and xylose as an internal standard for neutral sugars. The mass spectra of the aldononitrile acetates of muramic acid and N-methylglucamine show prominent fragments at mass to charge ratio, m/e, 115 and 86, respectively, which may be employed to monitor muramic acid and the internal standard, respectively, as shown in Figure 5-5 for Legionella pneumophila Philadelphia 1 [24]. The ratio of the muramic acid peak area to the internal standard area is used to predict the amount of muramic acid present in the organism by comparison to an external standard calibration in which known amounts of muramic acid

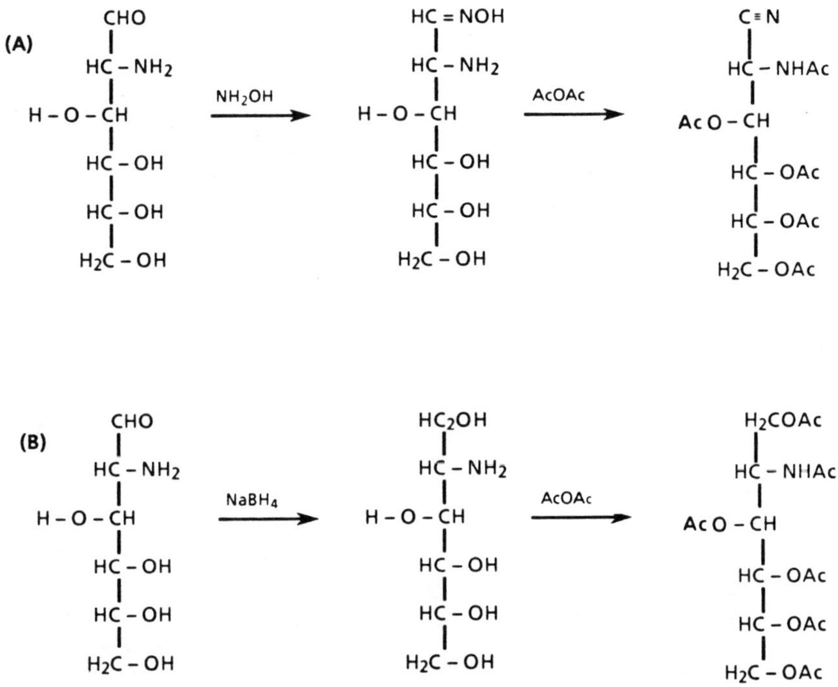

(A)

$$
\begin{array}{ccccc}
\text{CHO} & & \text{HC=NOH} & & \text{C≡N} \\
\text{HC-NH}_2 & & \text{HC-NH}_2 & & \text{HC-NHAc} \\
\text{H-O-CH} & \xrightarrow{\text{NH}_2\text{OH}} & \text{H-O-CH} & \xrightarrow{\text{AcOAc}} & \text{AcO-CH} \\
\text{HC-OH} & & \text{HC-OH} & & \text{HC-OAc} \\
\text{HC-OH} & & \text{HC-OH} & & \text{HC-OAc} \\
\text{H}_2\text{C-OH} & & \text{H}_2\text{C-OH} & & \text{H}_2\text{C-OAc}
\end{array}
$$

(B)

$$
\begin{array}{ccccc}
\text{CHO} & & \text{HC}_2\text{OH} & & \text{H}_2\text{COAc} \\
\text{HC-NH}_2 & & \text{HC-NH}_2 & & \text{HC-NHAc} \\
\text{H-O-CH} & \xrightarrow{\text{NaBH}_4} & \text{H-O-CH} & \xrightarrow{\text{AcOAc}} & \text{AcO-CH} \\
\text{HC-OH} & & \text{HC-OH} & & \text{HC-OAc} \\
\text{HC-OH} & & \text{HC-OH} & & \text{HC-OAc} \\
\text{H}_2\text{C-OH} & & \text{H}_2\text{C-OH} & & \text{H}_2\text{C-OAc}
\end{array}
$$

Figure 5-4. (A) The alditol acetate derivatization reaction and (B) aldononitrile acetate derivatization reaction.

and fixed amounts of N-methylglucamine are derivatized. The muramic acid contents determined for 22 bacterial strains are given by Table 5-3. Gram-positive bacteria were found to contain higher amounts of muramic acid than Gram-negative bacteria. Moriarty [27] has reported similar results. Among the 22 strains analyzed, we included eight legionella strains whose Gram type, although reported to be negative [28-30], is still under investigation. The amount of muramic acid in Legionella was within the lower range for Gram-positive organisms or within the range of Gram-negative organisms.

Other markers could be analyzed to supplement muramic acid determinations as a measure of Gram type. Figure 5-6 shows high-resolution capillary chromatograms of alditol acetates from whole-cell samples of two different organisms: (A) a Gram-positive organism, Streptococcus pyogenes, and (B) a Gram-negative organism, Escherichia coli [23]. Component 13, heptose, appears only in the chromatogram of the Gram-negative organism and mass spectral comparison to a heptose standard confirms its identify (Figure 5-7). Although, heptose has been detected previously in purified LPS from Gram-negative bacteria, the sensitivity of GCMS for trace detection of bacterial components is illustrated well by the detection of heptose in whole bacterial cells.

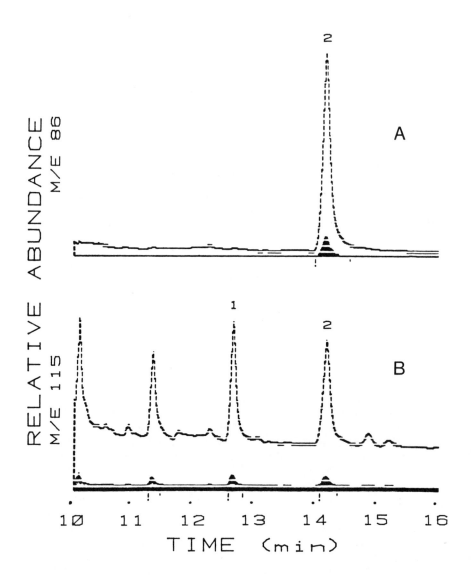

Figure 5-5. Selected ion monitoring chromatograms of L. pneumophila Phila-delphia 1: (A) m/e 86, and (B) m/e 115. Peak identification: (1) muramic acid, and (2) N-methylglucamine. (From ref. 24. Reprinted with permission of The Royal Society of Chemistry.)

Table 5-3. Results of Muramic Acid Determinations for Gram-positive,
Gram-negative, and Legionella Bacteria.[a]

Bacteria		Percentage Dry Weight
Gram-positive bacteria		
A	M. lysodeikticus	3.9
B	S. aureus	2.9
C	S. epidermidis	1.9
D	B. globigii	0.56
E	S. pyogenes	0.53
F	B. subtilis	0.48
G	C. perfringens	0.29
H	P. acnes	0.16
Gram-negative bacteria		
I	A. vinelandii	0.38
J	E. cloacae	0.29
K	A. aerogenes	0.29
L	K. pneumoniae	0.26
M	P. fluorescens	0.18
N	E. coli	0.17
O	P. aeruginosa	0.14
P	P. vulgaris	0.14
Legionellaceae		
Q	F. unclassified E-327F	0.30
R	L. pneumophila Knoxville	0.29
S	T. micdadei PPA-EK	0.28
T	T. micdadei PPA-GL	0.25
U	F. dumoffii NY-23	0.20
V	L. pneumophila Philadelphia	0.19

[a]From ref. 24. Reprinted with permission of The Royal Society of Chemistry.

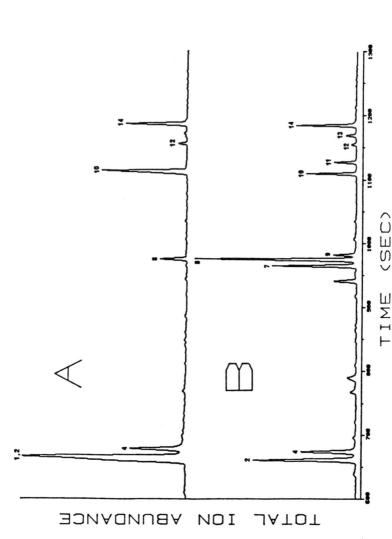

Figure 5-6. Reconstructed total ion abundance chromatograms of the alditol acetates from (A) S. pyogenes and (B) E. coli. Peak identities are: 1, rhamnose; 2, ribose; 4, arabinose (Neutral sugar internal standard); 7, mannose; 8, glucose; 9, galactose; 10, glucosamine; 11, galactosamine; 12, muramic acid; 13, heptose; 14, methylglucamine (amino sugar internal standard). (From ref. 23. Reprinted with permission of Elsevier Science Publishers, B.V.)

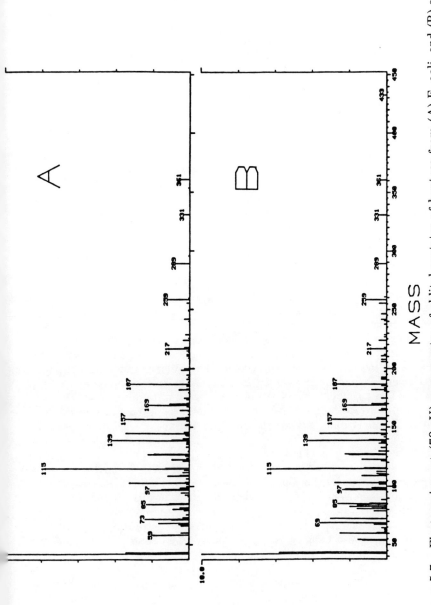

Figure 5-7. Electron impact (70 eV) mass spectra of alditol acetates of heptose from (A) E. coli, and (B) a heptose standard. The heights of peaks in the mass spectra are multiplied by 10. (From ref. 23. Reprinted with permission of Elsevier Science Publishers, B.V.)

IV. APPLICATION OF PYROLYSIS GCMS TO DETERMINING
GRAM TYPE OF BACTERIA

Although analytical pyrolysis has been applied to the differentiation of microorganisms through simple fingerprinting techniques, the chemical identification of thermal fragments and of the structural origin of these pyrolyzates within bacterial cells generally has not been done. Among the initial studies reporting systematic identification of pyrolyzates was that of Simmonds and co-workers [31,32], who attempted to correlate the formation of specific pyrolyzates with their site of origin in the structure of microorganisms. Their assignment of pyrolysis fragments to possible biological sources (such as protein, carbohydrate, nucleic acid, lipid, and porphyrin) is shown in Table 5-4. The wealth of pyrolyzates generated from bacteria provide a tremendous potential for differentiating subtleties of bacterial structure. Figure 5-8 presents high-resolution capillary pyrograms of three bacteria. The two Gram-negative bacteria (A and B) generate higher levels of furfuryl alcohol in comparison to the Gram-positive organism, which generates higher levels of acetamide upon pyrolysis. Simmonds [31]speculated that acetamide is produced from the N-acetyl groups of peptidoglycan. Model compounds with chemical structures similar to substructures within peptidoglycan have been investigated with pyrolysis GCMS by Hudson et al [33] and by Eudy et al [34]. Low levels of acetamide were found to be present in the pyrograms of glucosamine, muramic acid, and trialanine, but N-acetylated compounds (N-acetyl glucosamine, N-acetylmuramic acid, and muramyl dipeptide) produced much larger amounts of acetamide on pyrolysis. Although these results do not rule out other acetamide sources within microorganisms, N-acetyl groups on the glycan backbone do appear to be a major source of acetamide in pyrograms of peptidoglycan. The other prominent pyrolyzate, furfuryl alcohol, had been assigned by Simmonds to a carbohydrate origin. Eudy et al [34] found furfuryl alcohol to be a major pyrolyzate of DNA and RNA and performed comparative experiments on several model compounds, including 2-deoxyribose, ribose, glucose, and rhamnose. The deoxypentose, 2-deoxyribose, produced almost 20 times as much furfuryl alcohol as ribose and 100 times more than glucose or rhamnose. These results are not surprising when viewed in terms of the structural similarity of 2-deoxyribose to the furfuryl alcohol pyrolyzate. Just as the amount of peptidoglycan present in microorganisms does not fully account for the amount of acetamide produced as a pyrolyzate, likewise, the amount of DNA present in microorganisms does not account for all the furfuryl alcohol generated by pyrolysis. Both these pyrolyzates must also be generated from many other sources in the structure of the microorganism.

The discussion of acetamide and furfuryl alcohol pyrolyzates could be followed by similar discussions on the origin of other pyrolyzates in bacterial pyrograms. In many cases, similar studies with model compounds have not yet been performed. Nevertheless, the above-mentioned studies do point out the potential for directly correlating the production of certain pyrolyzates with specific bacterial components. Bacterial pyrograms are exceedingly complex,

Table 5-4. Assignment of Pyrolysis Fragments to Biological Classes
of Origin.[a]

Protein

Acetonitrile
Acrylonitrile
Benzene
Benzonitrile
Butanenitrile
Butene
C3 Alkyl benzene
C8 Alkyl benzene
C3 Alkyl pyrroles
C4 Alkyl pyrroles
Dimethyl pyridine
Dimethyl pyrroles
Ethan
Ethanenitrile
Ethene
Ethyl benzene
Ethyl phenol
Ethylene oxide
Indole
Isobutene
Isobutyronitrile
Isocapronitrile
Isopentene
Isovaleronitrile
Methane
Methanethiol
2-Methyl propanenitrile
Methyl butanenitrile
Methyl butene
Methyl indole
Methyl pentanenitrile
Methyl propene
Methyl pyridine
Methyl pyrroles
o-Cresol
p-Cresol
Phenol
Phenyl acetonitrile
Propane
Propanenitrile
Propene
Propyl benzene

m-Xylene
o-Xylene
p-Xylene
Xylenols

Carbohydrate

Acetone
Acrolein
Benzene
2-Butanone
Cyclopentadiene
Dimethyl furan
Furan
Furfural
Furfuryl alcohol
Isobutyraldehyde
Isovaleraldehyde
Methyl butadiene
Methyl butanal
Methyl cyclopentadiene
5-Methyl 2-furfural
Methyl propanal
2-Methyl furan
2-Pentanone
Propanal

Nucleic acid

Acetonitrile
Acrylonitrile
Butanenitrile
Ethanenitrile
Methyl pyridine
Propanenitrile
Pyridine

Lipid

Acrolein
Butene
Ethene
Propene

Table 5-4. Assignment of Pyrolysis Fragments to Biological Classes
of Origin.[a]
(continued)

Protein (continued)
   Pyridine   Porphyrin
   Pyrrole
   Styrene   C_3 Alkyl pyrroles
   Toluene   C_4 Alkyl pyrroles
   Tolunitrile   Dimethyl pyrroles
                         Methyl pyrroles
                         Pyrrole

[a]Adapted from ref. [31] and [32].

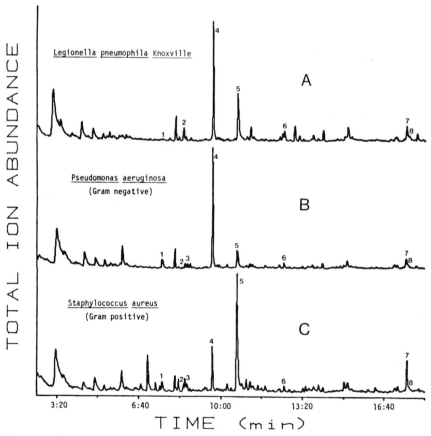

Figure 5-8.   Total ion abundance pyrograms of (A) Legionella pneumophila
Knoxville, (B) P. aeruginosa, and (C) S. aureus. Peak identifications given in
Table 5-5. (From ref. 24. Reprinted with permission of The Royal Society of
Chemistry.)

often containing several hundred peaks. Eudy et al [24] employed selected monitoring (SIM) to quantitate the peak areas of eight pyrolyzates (Table 5-5) from a diverse group of bacteria (Table 5-3). The use of SIM simplified the resulting pyrogram and reduced the amount of data to be handled. Because the comparison of pyrograms becomes unwieldy when there are a large number of them, and when each pyrogram has a large number of pyrolyzates, the data were treated by computer-assisted cluster analysis techniques. Each pyrogram with eight pyrolyzate intensities can be represented as a point in an eight-dimensional euclidean space. Nonlinear mapping then can be employed to reduce the multidimensional data to a two-dimensional display that approximates the structure and distances between points in the higher dimensional space [35,36]. Figure 5-9 presents a nonlinear map of the 42 pyrograms of this set of bacteria. Data points representing replicate measurements are connected by lines, the relative length of which is indicative of the reproducibility. The relative standard deviation for peak areas in replicate experiments was typically less than 10%. The nonlinear map shows that differentiation between Gram-positive and Gram-negative organisms is possible. Gram-positive bacteria lie in the upper half of the map, whereas Gram-negative bacteria lie in the bottom half of the map. The Legionella data points cluster in a region of the nonlinear map overlapping with one of the Gram-negative bacteria (Pseudomonas aeruginosa). These results agree with morphological and biochemical studies indicating that legionellae are Gram-negative bacteria.

Table 5-5. Ions Monitored, Retention Times, and Tentative Pyrolyzate Identification for Pyrolysis GCMS Study.[a]

Peak	Ion	Retention Time	Pyrolyzate Identification
1	54	7.6	2-Furancarboxaldehyde
2	67'	8.6	Unknown
3	98'	8.7	Unknown
4	98	9.7	Furfuryl alcohol
5	59	10.7	Acetamide
6	117	12.6	Benzeneacetonitrile
7	117'	17.6	1H-indole
8	99	17.9	2,5-Pyrrolidinedione

[a]From ref. 24. Reprinted with the permission of The Royal Society of Chemistry.

V. DIFFERENTIATION OF BACTERIAL SPECIES BY
CAPILLARY GC AND GCMS

Capillary GC can be a useful tool for the differentiation of bacterial species whenever specific chemicals exist that are characteristic of certain groups of microorganisms. As mentioned previously, muramic acid is one possible chemical marker whose detection implies the presence of bacteria. Other compounds, including heptose and KDO, are specific for Gram-negative bacteria. Besides the possibility of such chemical components being specific indicators of a particular class of microorganism, the relative amounts of common components (neutral sugars, such as rhamnose or fucose, or perhaps amino sugars, such as glucosamine) also may permit differentiation. The analysis of these components also provides fundamental chemical structure information that may be employed to differentiate or otherwise classify microorganisms.

A number of investigators have employed derivatization prior to GC to profile structural components of microorganisms. Moss and co-workers [37,39] used fatty acid profiling to distinguish bacterial species, including members of

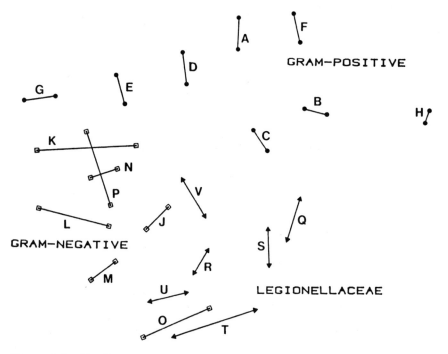

Figure 5-9. Nonlinear map of a group of pyrograms from Gram-positive and Gram-negative bacteria (listed in Table 5-3, excluding A. vinelandii). Points representing replicate measurements are connected by lines. ●, Gram positive; □, Gram negative; ▲, Legionellaceae.

the family Legionellaceae. Mayberry [40] extended these profiling techniques to bacterial hydroxylated fatty acids. Both sets of researchers employed the relatively nonselective flame ionization detector. Pritchard et al [41] , however, have performed carbohydrate profiling with a selective electron capture detector to differentiate streptococci. There have been surprisingly few attempts to apply GCMS to bacterial characterization [2,42,43]. Pyrolysis GC and GCMS also have been used to characterize microorganisms, and progress in this area has been reviewed recently by Bayer and Morgan [44] and others [2,3,45]. Gutteridge and Norris [46] have differentiated a number of Gram-negative and Gram-positive bacteria and French et al [47] have used isothermal pyrolysis GC with packed columns to identify streptococci. Abbey et al [48] recently have employed pyrolysis GCMS to characterize different strains of Klebsiella. Although techniques for the differentiation of bacteria by structural analyses using GC or GCMS have been successful in specific research laboratories, these techniques (either derivatization or pyrolysis) have not as yet been widely accepted for routine laboratory applications. One of the few methods that are used routinely is the identification of anaerobic bacteria by GC analysis of volatile metabolic products [49]. Even this technique, however, is performed manually, although Larsson [50] has automated the procedure recently with headspace analysis.

Work in our laboratories has focused on the development of improved methods for the analysis of carbohydrates in bacterial samples using a modified alditol acetate method [22,23,51,42]. The principal advantage of the alditol acetate method (Figure 5-4A) over other procedures, such as trimethyl-silyation, is that a single derivative is produced for each component sugar. This simplicity may be important in the analysis of a mixture of sugars from a biological matrix in which the production of multiple peaks for each component sugar gives a more complicated chromatogram that may confound qualitative identification and quantitative measurement. Another aspect of this method that should be stressed is the importance of sample cleanup prior to GC or GCMS analysis [22]. Following hydrolysis of the bacterial sample to produce monomeric sugars and neutralization of the excess acid, disposable hydrophobic columns are employed to remove lipids released by the hydrolysis as well as any residual reagents. Following the acetylation reaction, disposable hydrophilic columns are employed to remove the acetylation catalyst (sodium acetate) and other potential polar contaminants (such as amino acids). These "cleanup" steps coupled with other modifications to the alditol acetate method result in a chromatogram that is essentially free of extraneous peaks that may affect the quality of the chromatographic analysis adversely. Deactivated glass capillary columns coated with SP-2330 [51] or fused silica capillary columns coated with SE-52 and crosslinked for high-temperature stability [23] provide high-resolution separation and excellent sensitivity for mixtures of both neutral and amino sugars.

We have employed capillary GC [51] and GCMS [23] for the profiling of neutral and amino sugars present in several legionellae and other bacteria. Figure 5-10 demonstrates that carbohydrate profiling can differentiate three different groups of the Legionellaceae: L. pneumophila, Tatlockia micdadei,

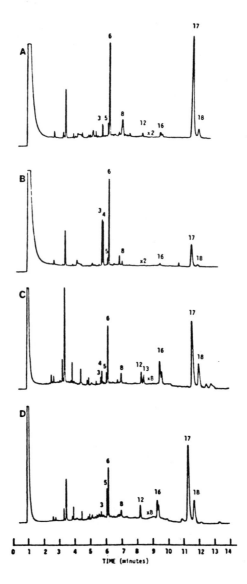

Figure 5-10. Capillary column chromatograms of whole bacterial cell hydrolyzates from (A) L. pneumophila, (B) T. micdadei, (C) F. bozemanae, (D) F. dumoffii. When chromatograms are compared it should be noted that there were three times the amount of internal standards in samples A and B compared to samples C and D. Glucose was also present in trace amounts in these organisms as confirmed by GCMS. Peak identification: 3, rhamnose; 4, fucose; 5, ribose; 6, arabinose; 7, xylose; 8, mannose; 12, aminodideoxyhexose XI; 13, aminodideoxyhexose X2; 16, muramic acid; 17, N-methylglucamine; 18, glucosamine. (From ref. 52. Reprinted with permission of the American Society for Microbiology.)

and Fluoribacter species. Legionella pneumophila was characterized by the
absence of fucose and by the presence of an amino dideoxyhexose. Fluoribacter strains were quite variable in their sugar composition, with four out of
five strains containing moderate amounts of rhamnose, and with three of five
containing moderate amounts of fucose. Reconstructed ion plots were employed effectively to confirm the presence of mannose, glucose, and galactose
in hydrolyzates from Legionella [23]. Although mannose was detected easily
by capillary GC in all samples [52], the mannose and galactose peaks were
quite small relative to the baseline noise, which made identification difficult
with flame ionization detection or with total ion abundance GCMS. As seen
in Figure 5-11, plotting selected characteristic masses (in this case, mass 115)
helps to distinguish these peaks from the background of other material
eluting at the same retention time.

Muramic acid and glucosamine were detected in all Legionella samples.
The amounts of glucosamine, however, were generally higher than the
amounts of muramic acid, possibly indicating the presence of glucosamine elsewhere in the bacteria. The presence of two amino dideoxyhexose isomers also

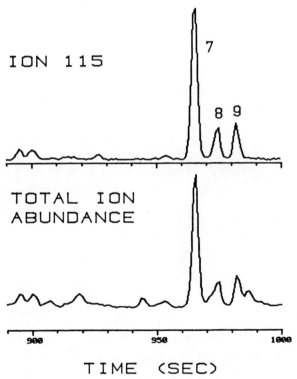

Figure 5-11. Reconstructed ion (mass 115) and total ion (mass 40-450) plots
of a portion of a chromatogram of alditol acetates from L. pneumophila showing the presence of mannose (peak 7), glucose (peak 8), and galactose (peak 9).
(From ref. 23. Reprinted with permission of Elsevier Science Publishers, B.V.)

was noted in four of the five <u>Fluoribacter</u> strains [23]. <u>Fluoribacter</u> strain NY-23 contained only one of these amino dideoxyhexose isomers. <u>Tatlockia micdadei</u> was distinguished by the absence of either amino dideoxyhexose and by the presence of large amounts of fucose and rhamnose. The two isomer peaks appearing in the chromatograms from <u>Fluoribacter</u> strains (peaks 12 and 13 in Figure 5-10C were identified as alditol acetates of amino dideoxyhexoses by electron impact and chemical ionization (CI) mass spectra. Electron impact mass spectrometry did not provide a molecular ion indicating the correct molecular weight of these two components. Methane CI mass spectra of one of the amino dideoxyhexoses and of glucosamine from <u>F. bozemanae</u> are shown in Figure 5-12. The molecular weight of the amino dideoxyhexose may be assigned as 375 based on interpreting the observed mass at 376 as the addition of a proton (M + 1 peak). The loss of acetic acid (M + H − 60) and of ketene M + H − 42) are also observed at mass 316 and

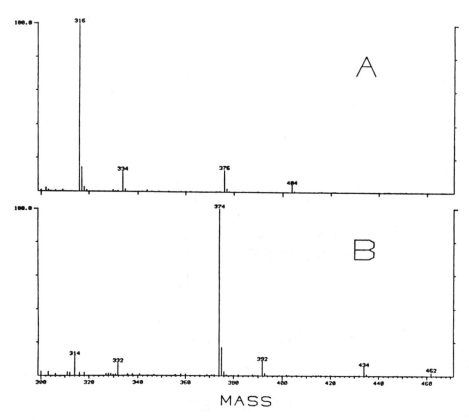

<u>Figure 5-12</u>. A comparison of the methane chemical ionization mass spectra of alditol acetates of sugars (A) amino dideoxyhexose and (B) glucosamine from <u>F. bozemanae</u>. (From ref. 23. Reprinted with permission of Elsevier Science Publishers, B.V.)

334, respectively. The peak at mass 404 results from the addition of C_2H_5 (M + 29).

We have also employed pyrolysis GCMS with SIM to differentiate members of the family Legionellaceae [53]. The organisms employed in this study are listed in Table 5-6. Replicate samples of each of the 21 organisms were pyrolyzed and 15 pyrolyzates were monitored by SIM GCMS at 10 different masses. The pyrolyzates included acetamide (mass 59), pyrrole (mass 67), pyridine (mass 79), furfuryl alcohol (mass 98), cresol (mass 108), benzeneace-tonitrile (mass 117), 1H-indole (mass 117 at a different retention time), and eight additional pyrolyzates monitored at masses 98, 54, 78, 94, 99, 99, 54, and 67. The set of 42 data points in a 15-dimensional space were normalized and autoscaled, and a two-dimensional nonlinear map of the data was plotted

Table 5-6. Listing of Legionellaceae Pyrolyzed

Legionella pneumophila

A Philadelphia
B Knoxville
C Pontiac
D SCH
E Togus 1
F Bloomington
G Los Angeles
H 684
I Houston

Tatlockia micdadei

J PPA-EK
K PPA-PGH-12
L PPA-JC
M PPA-ML
N PPA-GL
O PPA-CAR
P P. Tatlock

Fluoribacter bozemanae

Q Wiga
R MI-15
S NY-23 (dumoffi)
T LS-13 (gormanii)
U Unclassified E-327F

(Figure 5-13). Areas of the map occupied by the Legionella (samples A-I), Tatlockia (samples J-P), and Fluoribacter (sampels Q-U) groups are indicated. Legionella and Tatlockia groups are well separated; the Fluoribacter samples appear in an intermediate region and overlap slightly with both of the other groups. The partial separation of these groups of Legionella organisms indicates the potential of pyrolysis GCMS for bacterial differentiation. Certainly, the potential for bacterial characterization illustrated here supports previous work in the analytical pyrolysis literature [2,3,44].

VI. TRACE DETECTION OF CHEMICAL MARKERS FOR BACTERIAL PATHOGENS IN INFECTED TISSUES AND BODY FLUIDS

As a logical extension to the chemotaxonomic characterization of microorganisms, it has been suggested that it may prove possible to employ SIM GCMS in the direct diagnosis of bacterial infections by analyzing body fluids or tissues for chemical markers derived from bacterial cells. Although the application of GCMS has been limited so far to just a few published reports [54], the concept holds great promise. Maitra et al [55] have detected β-hydroxymyristic acid in serum to which small amounts of endotoxin had

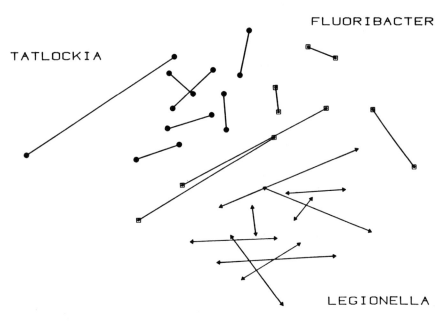

Figure 5-13. Nonlinear map of the pyrograms from a group of bacteria from the family Legionellaceae. Points representing replicate measurements are connected by lines. ●, Tatlockia; □, Fluoribacter; ▲, Legionella.

detection of chemical markers in infected body fluids and tissues to definitively indicate the presence of pathogenic microorganisms.

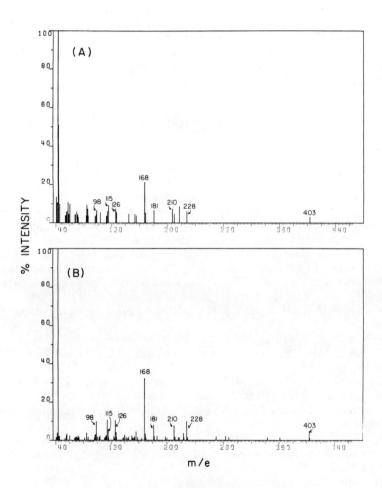

Figure 5-15. Electron ionization mass spectra of (A) the muramic acid peak in a chromatogram of a spleen extract from an animal injected 8 d previously with cell wall fragments; (B) pentaacetylmuramicitol reference standard. (From ref. 57. Reprinted with permission of the American Society for Microbiology.)

ACKNOWLEDGMENTS

This work was supported by NIH Grants GM-27135 (to S.L.M.) and EY-04715 (to A.F. and S.L.M.) and by the Veterans' Administration Medical Research Service. The Finnigan GCMS was purchased with funds from NSF through the EPSCOR Program and from the University of South Carolina. A.F. was also a Fellow of the Charles E. Culpeper Foundation while this review was in preparation. The authors gratefully acknowledge the GC and GCMS work of Michael D. Walla, Larry W. Eudy, and Pauline Y. Lau (Department of Chemistry, University of South Carolina). The assistance of Dr. Arnold Brown (Department of Medicine, School of Medicine, University of South Carolina) in supplying Legionella samples and the work of Matthew Przybyciel (Department of Chemistry, University of South Carolina) in preparation of capillary columns is also recognized.

VIII. REFERENCES

1. Drucker, D.B. "Microbiological Applications of Gas Chromatography"; Cambridge University Press: Cambridge, 1981.
2. Gutteridge, C.S.; Norris, J.R. J. Appl. Bacteriol., 1979, 47, 5.
3. Meuzelaar, H.L.C.; Haverkamp, J.; Hileman, F.D. "Pyrolysis Mass Spectrometry of Recent and Fossil Biomaterials: Compendium and Atlas"; Elsevier Scientific Publishers Co.: Amsterdam, 1982.
4. Larssen, L.; Mardh, P.-A.; Odham, G. Lab Management, 1983, 21, 38.
5. Buchanan, R.E.; Gibbons, N.E., eds. "Bergey's Manual of Determinative Bacteriology"; 8th ed.; The Williams and Wilkins Company: Baltimore, 1974.
6. Fieldsteel, A.H.; Cox, D.L.; Moeckli, R.A. Infect. Immun., 1981, 32, 908.
7. Brown, A. Lab World, 1981, 32, 38.
8. Brenner, D.J.; Steigerwalt, A.G.; McDade, J.E. Ann. Intern. Med., 1979, 90, 656.
9. Cordes, L.G.; Wilkinson, H.W.; Gorman, G.W.; Fikes, B.J.; Fraser, D.W. Lancet, 1979, 11, 927.
10. Garrity, G.M.; Brown, A.; Vickers, R.M. Int. J. Syst. Bacteriol., 1980, 30, 609.
11. Brooks, C.J.W.; Edmonds, C.G. In "Practical Mass Spectrometry: A Contemporary Introduction"; Middleditch, B.S., ed.; Plenum Press: New York, 1979.
12. Schleifer, K.H.; Kandler, O. Bacteriol. Rev., 1972, 36, 407.
13. Luderitz, O.; Galanos, C.; Lehman, V.; Nurminen, M.; Rietschel, E.T.; Rosenfelder, G.; Simon, M.; Westphal, O. J. Infect. Dis., 1973, 128, S17.
14. Inouye, M. "Bacterial Outer Membranes"; John Wiley and Sons: New York, 1979.
15. Corina, D.L.; Sesardic, D. J. Gen. Microbiol., 1980, 116, 61.

16. Wiegel, J. Int. J. Syst. Bacteriol., 1981, 31, 88.
17. Cerny, G. Eur. J. Appl. Microbiol., 1976, 3, 223.
18. Wiegel, J.; Quandt, L. J. Gen. Microbiol., 1982, 128, 2261.
19. Halebian, S.; Harris, B.; Finegold, S.M.; Rolfe, R.D. J. Clin. Microbiol., 1981, 13, 444.
20. Buck, J.D. Appl. Environ. Microbiol., 1982, 13, 992.
21. Carlone, G.M.; Valadez, M.J.; Pickett, M.J. J. Clin. Microbiol., 1983, 16, 1157.
22. Fox, A.; Morgan, S.L.; Hudson, J.R.; Zhu, Z.T.; Lau, P.Y. J. Chromatogr., 1982, 256, 429.
23. Walla, M.D.; Lau, P.Y.; Morgan, S.L.; Fox, A.; Brown, A. J. Chromatogr., 1984, 288, 399.
24. Eudy, L.W.; Walla, M.D.; Morgan, S.L.; Fox, A. Analyst, in press (1985).
25. Fazio, S.D.; Mayberry, W.R.; White, D.C. Appl. Environ. Microbiol., 1979, 38, 349.
26. Findlay, R.H.; Moriarty, D.J.W.; White, D.C. Geomicrobiol. J., 1983, 3, 135.
27. Moriarty, D.J.W. Oecologia (Berlin), 1977, 26, 317.
28. Keel, J.A.; Finnerty, W.R.; Feeley, J.C. Ann. Intern. Med., 1979, 90, 652.
29. Rogers, F.D.; Davey, M.R. J. Gen. Microbiol., 1982, 128, 1547.
30. Wong, K.H.; Moss, C.W.; Hochstein, D.H.; Arko, R.J.; Schalla, W.O. Ann. Intern. Med., 1979, 90, 624.
31. Simmonds, P.G. Appl. Microbiol., 1970, 20, 567.
32. Medley, E.E.; Simmonds, P.G.; Mannatt, S.L. Biomed. Mass Spectrom., 1975, 2, 261.
33. Hudson, J.R.; Morgan, S.L.; Fox, A. Anal. Biochem., 1982, 120, 59.
34. Eudy, L.W.; Walla, M.D.; Hudson, J.R.; Morgan, S.L.; Fox, A. J. Anal. Appl. Pyrol., in press (1985).
35. Eshuis, W.; Kistemaker, P.G.; Meuzelaar, H.L.C. In "Analytical Pyrolysis"; Jones, C.E.R.; Cramers, C.A., eds.; Elsevier Scientific Publishing Co.: Amsterdam, 1977, p. 156.
36. Morgan, S.L.; Jacques, C.A. Anal. Chem., 1982, 54, 741.
37. Moss, C.W.; Dees, S.B. J. Clin. Microbiol., 1979, 10, 390.
38. Moss, C.W.; Karr, D.E.; Dees, S.B. J. Clin. Microbiol., 1981, 14, 692.
39. Moss, C.W.; Weaver, R.E.; Dees, S.B.; Cherry, W.B. J. Clin. Microbiol., 1977, 6, 140.
40. Mayberry, W.R. J. Bacteriol., 1981, 147, 373.
41. Pritchard, D.G.; Coligan, T.E.; Speed, S.E.; Gray, B.M. J. Clin. Microbiol., 1981, 13, 89.
42. Alvin, C.; Larsson, L.; Magnusson, M.; Mardh, P.-A.; Odham, G.; Westerdahl, G. J. Gen. Microbiol., 1983, 129, 401.
43. Bergman, R.; Larsson, L.; Odham, G.; Westerdahl, G. J. Microbiol. Meth., 1983, 1, 19.
44. Bayer, F.L.; Morgan, S.L. In "Pyrolysis and GC in Polymer Analysis"; Liebeman, S.A.; Levy, E.J., eds.; Marcel Dekker: New York, 1985.
45. Irwin, W.J. "Analytical Pyrolysis: A Comprehensive Guide"; Marcel Dekker: New York, 1982.

46. Gutteridge, C.S.; Norris, J.R. Appl. Environ. Microbiol., 1980, 40, 462.
47. French, G.L.; Phillips, I.; Chinn, S. J. Gen. Microbiol., 1981, 125, 347.
48. Abbey, L.E.; Highsmith, A.K.; Moran, T.F.; Reiner, E.J. J. Clin. Microbiol., 1981, 13, 313.
49. Holdeman, L.V.; Cato, E.P.; Moore, W.E.G., eds. "Anaerobe Laboratory Manual", 4th ed.; Anaerobe Laboratory, Virginia Polytechnic Institute and State University: Blacksburg, VA, 1977.
50. Larsson, L.; Holst, E. Acta Pathol Microbiol. Scand., Sect. B, 1982, 90, 125.
51. Hudson, J.R.; Morgan, S.L.; Fox, A. J. High Resolut. Chromatogr.; Chromatogr. Commun., 1982, 5, 285.
52. Fox, A.; Lau, P.Y.; Brown, A.; Morgan, S.L.; Zhu, Z.T.; Lema, M.; Walla, M.D. J. Clin. Microbiol., 1984, 19, 326.
53. Eudy, L.W., Ph.D. Dissertation, University of South Carolina, Columbia, SC, 1983.
54. Larsson, L.; Mardh, P.-A.; Odham, G.; Westerdahl, G. Acta Pathol. Microbiol. Scand., Sect. B, 1981, 89, 245.
55. Maitra, S.K.; Schotz, M.C.; Yoshikawa, T.T.; Guze, L.B. Proc. Natl. Acad. Sci. U.S.A., 1978, 75, 3993.
56. Roboz, J.; Susuki, R.; Holland, J.F. J. Clin. Microbiol., 1980, 12, 594.
57. Fox, A.; Schwab, J.H.; Cochran, T. Infect. Immunol., 1980, 29, 526.
58. Sen, Z.; Karnovsky, M.L. Infect. Immunol., 1984, 43, 937.

6. ANALYSIS OF INDIVIDUAL BIOLOGICAL PARTICLES IN AIR

Mahadeva P. Sinha

I. INTRODUCTION

Monitoring air for biological contaminants has attracted renewed interest because of recent studies that have linked the spread of disease to the transfer of agents through an air medium [1]. Legionnaires' disease [2] typifies the air transfer of microbial agents and shows the difficulties in detecting and identifying unknown biological agents by conventional techniques.

Biological contaminants occur in air as aerosols, defined as solid or liquid particles suspended in air. These aerosols consist of bacteria, fungi, viruses, and pollen particles and cover a wide size range. The range of 1-10 μm is most important [30]. Particles larger than this, whether single cells or aggregates, do not remain suspended in air for a long time but settle on surfaces. They can multiply on congenial surfaces after settling and may again be made airborne by breaking as a result of external activities [4]. Microbial particles in air come from both artificially generated and naturally occurring sources: solid waste treatment [5], industries (eg, cotton milling [6], mushroom production [7]), wastewater treatment [8], irrigation projects [9], and bubbles bursting from microbially laden water layers [10], to name some sources. Furthermore, these particles do not always remain confined to the proximity of their sources. A typical 1-μm microbe can travel hundreds of kilometers downwind before settling and affecting the environment. Airborne contagions have been found to cause the spread of infections in hospital environments [11], where patients are likely to be more susceptible to disease. People disperse microbes while talking, coughing, and sneezing [12], and open wounds, skin diseases, and skin lesions [13] shed microbes that cause disease by air transfer. Analytical conditions under general field conditions would be more complex because of the interferences from substances normally present in air.

Obviously, it is desirable to monitor air for biological particles in a number of different environments; namely, hospitals, clean rooms, areas surrounding different industrial sites, and regions of military importance. The ideal method for such monitoring would be one that is fast, sensitive, and specific, but no single method possesses all of these features. Consequently, the choice of an analytical method depends on the circumstances. For example, speed may be the deciding criterion for an attack scenario, even at the expense of specificity (although a followup by a more specific method would have to be made in order to take the appropriate countermeasures).

The objectives of this chapter are (1) to briefly review the principles of several rapid methods of detection and identification of biological agents

165

(especially airborne ones), along with their speed, sensitivity, and specificity; and (2) to describe in detail a new technique known as particle analysis by mass spectrometry (PAMS) [14] for the direct analysis of individual particles on a real-time basis.

II. DETECTION OF BIOLOGICAL PARTICLES

The methods of detecting microorganisms in air are the same as in any other medium. These can be classified into methods based on the following:

1. Measuring physical characteristics of particles
2. Measuring biological activities
3. Detecting key biochemical substances
4. Generating characteristic fingerprint spectra

A. Physical Methods

The physical methods of detecting microorganisms depend on measuring the number, size, and shape of aerosol particles, and then the measurements indicate only the presence of aerosol particles in a suspected size range [15, 16]. For example, a sudden increase in the ratio of the number of particles in the 1-5 µm range to the total number of particles could be construed as an indication of artificially generated biological aerosol. Such an interpretation is based on the belief that the 1-5 µm size range is preferred for dispersion of these agents because of their importance to enhanced pulmonary retention and their prolonged suspension in air. So, physical measurements are rapid but are not specific and make no distinction between a microbial particle and a particle of any other origin. The importance of physical methods, however, lies in their speed (seconds).

Instruments originally developed for use in air pollution and industrial hygiene can be used for measuring microorganisms without any modification. Optical particle counters, aerosol photometers, nephelometers, and microscopes are used widely in bacteriology laboratories. The poor specificity for particles of biological origin can be improved by combining a physical method with a method from the other approaches listed above. A partichrome analyzer [15] is an example of such a combination. Here the particles are collected on a sticky tape and are stained with a dye. Biological particles retain the dye and can be examined with an automated microscope.

B. Biological Methods

The methods based on biological activity measure life processes. The microorganisms are studied during their growth by monitoring the increase in their mass or number and by monitoring the evolution of their metabolic products. Several instruments that measure such changes continuously have been built. The Gulliver device [17,18] uses a growth medium containing C^{14}-labeled sugar. During their growth, the microorganisms liberate $C^{14}O_2$, which is then measured by a Geiger-Muller counter. This particular method has the sensitivity to detect 10-100 cells. Measurements of the changes in turbidity and pH of the growth medium also have been used as indicators for biological activities. For example, the Wolf Trap [19] is a miniature growth chamber whose turbidity is continuously monitored by a nephelometer and whose pH changes are monitored by a pH electrode. Other instruments (eg, the multivator) [15,17] are based on the principle that certain media, on decomposition by microbial enzymes during growth, form chromogenic or fluorescent substances that can be detected by fluorimetric or colorimetric techniques.

Serological methods are highly specific and use antibody-antigen reactions [20] to detect and identify microorganisms. These methods apply to any microorganism for which an antibody may be available, and a number of antibodies can be produced and extracted from laboratory animals. The antibody-antigen reactions are so specific that even in a crude mixture they can be used for the rapid detection of a single antigen. The reactions can be accelerated, bringing the antibody and antigen together using electric current (immunoelectroosmophoreais [20]). Also, a fluorescent antibody can be prepared by linking a fluorescent dye, fluorescein, to the antibody globulin. After reaction the fluorescent antibody-coated organism stands out as a brilliantly fluorescent cell under a microscope with an ultraviolet light source [21-23]. Cultivation and isolation, however, usually are performed to confirm the diagnosis because one microorganism can give rise to several antibodies, some of which may be similar to those produced by other species.

The methods for detecting and identifying microorganisms based on their biological activities have the advantages of high specificity and sensitivity. A single cell can be detected, given enough time, and viable particles can be differentiated from nonviable particles. These methods, however, suffer from a lack of speed (the growth of organisms takes time, at least a few hours), and their application is not general because microbes have specific requirements. Furthermore, these methods necessitate collecting particles from air in the viable state. Also, although changes in turbidity and the pH of the medium normally are caused by living things, sometimes nonliving substances imitate characteristic viable activity. For example, crystals can grow in the medium, and some nonviable material can absorb or desorb gases [15], giving false indications of viability.

C. Biochemical Methods

Another class of methods depends on detecting biochemical components or unique chemical structures in microorganisms, chemical characteristics that distinguish biological from nonbiological materials. Commonly used for this purpose are tests for the presence of nucleic acids, ATP, and proteins in the biological particles. For example, biological compounds (eg, proteins, nucleic acids, peptides, carbohydrates) react with certain dyes, such as thiocarbocyanins [24,25], and produce a spectral shift in the light absorption characteristics of the dye. Measuring this shift by conventional spectroscopy has been the basis for detecting biological compounds in an unknown sample. Acridine orange [26], for instance, becomes bound to nucleic acids and produces fluorescence in the microbial particles under ultraviolet light. Another method (see Chapter 2) uses ATP, present in all living cells as the carrier of energy, to induce the reaction of the firefly enzyme luciferase with its substrate luciferine. Luciferase does not react with luciferine in the absence of ATP and so provides a test for ATP in a sample [27-30]. Another method of detection consists of heating the sample, which converts the protein in it to NH_4^+. The ions are then measured in an ion detection chamber [15]. This method is sensitive and can detect a sample as small as 0.1 µg of nebulized albumin, but it suffers from the noise generated by the presence of nonmicrobiological proteinaceous material in air.

The methods described above are fairly rapid and provide information about the biological nature of the sample, but they have several limitations. Notably, none of the methods requires viability of the sample. Consequently, these methods provide no information on the living state of the microorganisms and provide little information for differentiating among them. Furthermore, these methods cannot be used alone in field conditions for detecting biological particles because of possible interferences. For example, the method using acridine orange is compromised in the field by nonbiological particles that fluoresce either alone or in combination with the dye. Also, the complexities of the analysis of microorganisms in the field are increased because background material is present from natural sources, such as algae, pollen, and sawdust, and because a small number of nonpathogenic microorganisms usually exist in air. It is, therefore, important that the analytical methods used for detecting and identifying microbial particles in air not only should be sensitive to biological compounds but should also be specific enough to eliminate interferences.

Methods that use chromatography, mass spectrometry (MS), and various combinations of these two techniques meet such requirements (see Chapter 5) and have attracted much attention for the rapid analysis of microorganisms [31-35]. The composition of cell walls [36] and DNA content [37] differ markedly between bacterial species and therefore can be used for detection and identification. For the application of gas chromatographic and mass spectrometric methods, the analyte must be converted into the vapor phase. In addition, vaporization of the biological material must be preceded by fragmentation into smaller mass units. Two methods for the vaporization and fragmentation of large biological compounds have been used. In the first method, the macromolecules are converted into their small units by hydrolysis and subsequently are reacted with some suitable compounds to prepare their relatively volatile derivatives. The second method consists of heating a sample in a nonoxidizing atmosphere (pyrolysis) [38] whereby the large molecules are fragmented, producing small units in the vapor phase. The fragmentation takes place at preferred junctions in the molecules, and the fragments contain information about their parent molecules.

Morgan and Fox [39] recently have used derivatization gas chromatography (GC), high-resolution capillary GC, GCMS, and pyrolysis GCMS for bacterial analysis (Chapter 5). Specific compounds, for example, muramic acid, diaminopimelic acid, and D-amino acids, have been used as unique markers for the classification of bacteria. Gonser et al [40] studied the feasibility of an identification method based on mass spectrometric analysis of the nucleic acids bases of microorganisms. The method consists of extracting nucleic acids from microorganisms, hydrolyzing the acids to release purines and pyrimidines quantitatively, and volatilizing and analyzing the purine-pyrimidine mixture using field-ionization mass spectrometry. The success of the analytical method following pyrolysis largely depends on the pyrolysis conditions. It is important that the fragments produced by the bond scission not undergo any reaction among themselves or with the background gases. These secondary reactions tend to produce noncharacteristic fragments and to decrease the reproducibility of the pyrolysis process. It is preferable to use a method of pyrolysis that provides a controllable uniform temperature, a maximum heating rate, and a minimum heating time, and that uses a minimal amount of sample [41]. Reproducibility is enhanced if the entire sample experiences the same temperature, and if the pyrolysis products are exposed to a collision-free environment. The transit time from the pyrolysis site to the analyzer should be minimized in order to decrease the chances of pyrolyzates interacting among themselves or with the walls of the container.

Several designs of pyrolyzers have been reported in the literature and have been reviewed recently by Irwin [38] and Wieten et al [42]. They are classified in terms of their mode of operation; namely, the continuous mode (eg, furnace) and the pulse mode. The pulse-mode pyrolyzers commonly are used and include resistively heated pyrolyzers, Curie-point pyrolyzers, and laser pyrolyzers. Very encouraging results have been obtained with Curie-point pyrolyzers. In this method a small amount of sample in the range of 5-20 μg is deposited directly on a wire made of ferromagnetic material. High-frequency inductive heating is employed to raise the temperature of the wire to its Curie temperature. It takes about 100 ms to reach its equilibrium temperature, and typically the sample is pyrolyzed for a period of 1-10 s. Reliable and reproducible temperature-time profiles have been obtained. Meuzelaar and his co-workers have constructed fully automated systems for Curie-point pyrolysis-GC [43] and Curie-point pyrolysis-MS [44].

Although laser pyrolysis is still in its infancy, it possesses many desirable features [45]. Some of these are high temporal and spatial resolution, a very fast heating rate, and the ability to pyrolyze extremely small amounts of sample ($\sim 10^{-12}$ g) so that single-cell analysis is possible. A laser pyrolysis mass spectrometer is available commercially. The system, known as Laser Microprobe Mass Analyzer (LAMMA) [46,47], uses a frequency quadrupled Neodymium-YAG laser. The laser beam is focused to micron size and has a spatial resolution of this order. Applied to biological systems, the technique has been used to anlyze intracellular electrolytes [48] and trace metals with high sensitivity. The LAMMA also has been used to analyze single aerosol particles [49] deposited on glass substrates. The extent of the fragmentation of organic molecules is found to depend on the irradiance of the laser beam in focus. Also, laser desorption mass spectrometry [50,51] has been applied to the mass spectral measurements of nonvolatile bioorganic molecules. High mass fragments with good intensities are obtained, but the measurements seem to be qualitative in nature. Applicability of lasers to the pyrolysis of biopolymers and cells is yet to be demonstrated.

D. Fingerprint Spectral Methods

Pyrolysis gas-liquid chromatography (PyGLC) and pyrolysis-mass spectrometry (PyMS) have been used for fingerprinting complex biological material. Reiner [52] was able to distinguish visually the pyrograms of some of the closely related strains of microorganisms and pointed out the possibility of PyGLC forming a new basis of taxonomy in biology. Fingerprints of Salmonella bacteria were later matched by computer by Menger et al [53]. The main drawbacks of fingerprinting microorganisms by PyGLC are the lack of standardization and reproducibility of the results from different laboratories [54]. These deficiencies can be attributed to both the pyrolysis technique and the chromatographic columns: the difficulty in reproducing the pyrolysis conditions, the insufficient column resolution, the degradation of the column, and the chemical selection by the column. It may be possible to minimize these problems by using a well-controlled pyrolysis and by applying MS directly to the pyrolyzates. Direct application of MS to the pyrolysis product to obtain PyMS fingerprints of albumin and pepsin was reported by Zemany in 1952 [55]. Pyrolysis was performed by heating the proteins on a filament in a flask; no data on the reproducibility of the procedure were given. Recently, the combination of Curie-point pyrolysis with MS (using low-energy electron-impact ionization) has provided very promising results for the detection and identification of microorganisms [34]. High reproducibility in the pyrolysis pattern has been observed [56], and the low-voltage electron-impaction ionization has helped to retain the chemical information on the building blocks of the biomacromolecules. Also, PyMS has been shown to possess high sensitivity, speed, and computer compatibility. The addition of pattern recognition programs for mass spectral analysis to PyMS enhances specificity and speed. These advanced data analysis programs also help in identifying the chemical nature and the origin of mass peaks in a spectrum. The method has been applied successfully in the laboratory by Meuzelaar and others to characterize a wide range of biological materials, including bacteria, viruses, mammalian cells and tissues, body fluids, biopolymers, plant tissues, and other substances [57]. The detection and identification of microorganisms by comparing their fingerprint spectra require a compilation of standard fingerprints, so a library should be generated. An atlas of the reference spectra from complex biological compounds has been made by Meuzelaar et al [57]. A recent review of analytical pyrolysis in clinical and pharmaceutical microbiology has been made by Wieten et al [42]. The use of the advanced statistical analysis programs [42] for mass spectral data analysis has the potential of making PyMS a general method for biomaterial analysis.

However, all the above methods require the collection of a sample from air and its preparation in a suitable form before analysis. Methods of sampling airborne bacteria are in general the same as those used for any other particulate matter (eg, pollution aerosols). Commonly used methods of collection include impaction on solid surfaces (single- or multiple-stage impactors), impingement in liquids, and filteration [58]. The drawbacks of these sampling methods are that some preanalysis time is required and that the collection of all particles without any discrimination may introduce background noise in the analytical procedure. Because particles other than suspected pathogens are, in most environments, numerous in air, a desirable way to analyze the airborne microbial particles in the midst of a host of other particles would be to work on individual particles. This would provide a major improvement in the signal to noise ratio for the bioparticle analysis. In some circumstances; for example, in biological warfare situations, the speed of analysis is critically important.

Continuous, real-time monitoring of air samples for the presence of microorganisms, importantly, would eliminate sample collection and preparation phases. Such a method has been reported by Sinha et al [14]. Known as particle analysis by mass spectrometry (PAMS), the technique seems to hold promise for detecting, quantifying, and identifying airborne microorganisms.

III. PARTICLE ANALYSIS BY MASS SPECTROMETRY

Particle analysis by mass spectrometry (PAMS) may be classified as a PyMS technique with unique features: the direct introduction of aerosol particles in the form of a beam into the ion source of the mass spectrometer; the volatilization and ionization of one particle at a time under optimum conditions for pyrolysis; and the mass spectral measurement of individual particles. These features enable the analysis of single particles on a continuous, real-time basis.

The first direct introduction of airborne particles into a mass spectrometer was made by Meyers and Fite [59]. The work was extended by Davis [60] to analyze mainly the elemental composition of inorganic components of aerosol particles. Thermal ionization was used for the generation of ions, a method sensitive to elements with low ionization potential. Some organic amines also could be ionized by surface ionization [61]. However, no systematic study of the introduction of particles into the source was made, and the technique was limited to substances with low ionization potential. A more detailed study of the introduction of particles into the ion source of a mass spectrometer along with the different methods of ionization has since been made by Sinha et al [14] and Estes et al [62]. The method of electron-impact ionization is used, which makes the technique generally applicable. Preliminary studies on laser-induced volatilization and ionization of the particles in the beam has also been made [63, 64].

A. Particle Beams

Particle beams, analogous to molecular beams, are produced when an aerosol expands through a nozzle into a vacuum. These beams are highly directed and forward peaked and provide a well-controlled means of sample introduction into the mass spectrometer. Particle beams were first described by Murphy and Sears [65] and later studied by Israel and Friedlander [66] and by Dahneke and Friedlander [67] for their physical characteristics. The properties and applications of particle beams have been reviewed recently by Dahneke [68];

The main components of the PAMS system are the particle beam generator and a quadrupole mass spectrometer housed in differentially pumped chambers (Figure 6-1). A beam of particles is produced by an aerosol expanding into a

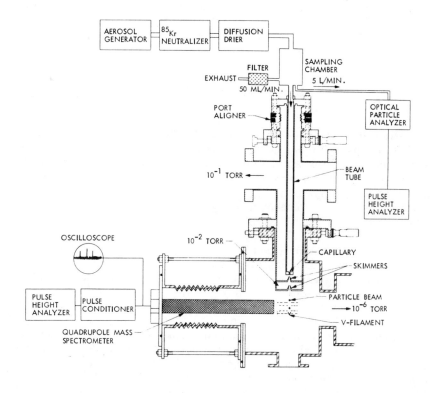

Figure 6-1. Schematic of PAMS systems. (Reproduced with permission from ref. 14.)

vacuum through the generator that consists of a capillary (100 μm in diameter and 5 mm in length) and a set of two skimmers with orifice diameters of 350 μm and 500 μm. The particles, being much heavier than the gas molecules, remain concentrated along the central axis. The skimmers allow an efficient transmission of the particles into the mass spectrometer chamber and provide for differential pumping of most of the carrier gas upstream. [The details of the PAMS system have been described elsewhere (14,69)]. Because the particles are massive compared to the gas molecules, they tend to be lost along the sample delivery line and physical losses occur on the surface because of settling, conventive flow, or diffusion [70]. The predominance of any one of these mechanisms depends upon the particle size and the flow velocity.

The particle delivery system should be designed to minimize the line loss of the particles, and it is important to characterize its transmission efficiency experimentally for particles of different sizes. For particles of diameters greater than 0.5 μm, the gravitational settling has the largest contribution to the particle loss [71]. This loss was eliminated in the PAMS system by directing the beam vertically downward. The transmission efficiency--the ratio of the number of particles in a given volume of the aerosol to the number of particles present in the beam generated from that volume--was determined by making beams of various monodisperse aerosols. This determination was made by measuring the number density of particles and the volume flow rate of the aerosol at the input of the beam tube and by measuring the number of particles entering the mass spectrometer chamber in a given time interval. The physical properties of particle beams under experimental conditions similar to those for PAMS also have been made by Estes et al [62]; the transmission efficiency measured is shown in Figure 6-2.

The other important parameter of a particle beam is its divergence. The angle of divergence may be measured by collecting the particles from the beam on a glass slide coated with either Vaseline or Apiezon grease. Measuring the deposit size and the distance of the glass slide from the capillary provides the angle of divergence. Such measurements need to be made in order to insure that the aerosol sample is not significantly modified in the course of its delivery to the mass spectrometer. Estes et al [62] have made such measurements for various parameters of the beam generator and for various particle sizes. They found that the particle beam generator has high transmission efficiencies (80%) for particles in the size range of 0.3-5 μm and that the divergence of the beam does not change drastically between particle diameters in the range of 1-5 μm. However, particles of smaller sizes are better focused along the beam axis. The transmission efficiencies of the beam generator and the beam divergence are particularly important when the beam is generated from a polydisperse aerosol sample (natural aerosol), and only a part of the cross section of the beam is analyzed.

B. Volatilization and Ionization of the Particles

The particle beam enters the mass spectrometer chamber through the skimmers and impinges on a V-shaped zone refined rhenium filament (Figure 6-3) located between the grid and the repeller of the ion source of the quadrupole mass spectrometer (Uthe Technology, Inc., Model 100C). The filament is heated resistively and maintained at a constant temperature in the range of 200-1400°C. A particle striking the filament produces a plume of vapor molecules; these are ionized by electron impaction in situ. Volatilization and ionization of each particle is a discrete event in time and results in a burst of ions for each particle, as shown in Figure 6-4. The ion pulse width has been found to be about 100-200 μs. The brevity of the pulse does not allow scanning for different ion masses.

C. Data Acquisition

A quadrupole mass spectrometer measures the intensities of different masses by scanning them in time. Because the available scan speed is too slow

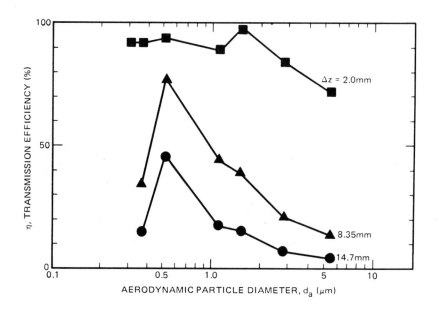

Figure 6-2. Overall particle transmission efficiency as a function of capillary skimmer separation (Δz) and particle aerodynamic diameter (d_a) where $d_a = \sqrt{\rho} \cdot d_g$, d_g is the geometric particle diameter, and ρ is the density of the particle. For larger Δz, the skimmer opening becomes the limiting aperture that decreases the transmission. In the case of PAMS, $\Delta z = 3$ mm. (Reproduced with permission from ref. 62, after Estes.)

for a complete mass analysis of a single ion pulse, only one mass peak is monitored from each particle. The mass range is scanned manually, and the intensity measurement is made wherever the signal is observed. The scheme for average intensity measurement is shown in Figure 6-5. Measuring the average intensity of a mass peak is particularly important when the input aerosol is polydisperse. The ion pulse at a particular mass is fed into a charge integrator that integrates the signal pulse for a preset time interval. Immediately after this, the background noise is integrated for the same length of time and subtracted from the stored signal. The distribution of the peak voltages of the integrated signal pulses is then measured with a pulse-height analyzer. The distribution is sufficiently narrow for the monodisperse aerosol and shows the reproducibility of the volatilization process. The peak of the distribution measures the average intensity of the particular mass peak. For a polydisperse aerosol, the intensities in various channels of a distribution are weighted by their respective number of particles. The sum of the number weighted intensities is then divided by the total number of particles to obtain their weighted average intensity. At a selected mass, about 1000 pulses are processed to obtain a distribution of single intensities. The mass spectrum is a collection of the average intensities of individually measured mass peaks and represents the spectrum of an average individual particle. One such spectrum of 1.7-μm diameter potassium biphthalate aerosol is shown in Figure 6-6.

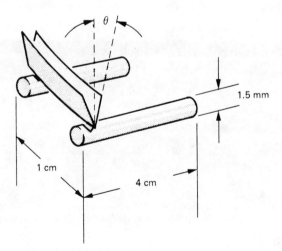

Figure 6-3. Design of Re V-type filament, $\theta \simeq 10^\circ$. The cylindrical elements support the filament and carry the heating current. (Reproduced with permission from ref. 14.)

IV. APPLICATION OF THE PAM TECHNIQUE

The PAMS technique has been applied to the mass spectral measurements of a variety of aerosols. These consist of aerosol particles composed of chemical species present in environmental pollution [14] (eg, ammonium salts, dibasic organic acids, lead salts, flyash [72]), aerosols generated from organic compounds [73] (which constitute the biological marcomolecules), and aerosols of microorganisms [69]. The aerosols are generated by various techniques in the laboratory. For compounds that are soluble in water or alcohol, a vibrating orifice aerosol generator [74] is used. This operates by forcing a

(a)

(b)

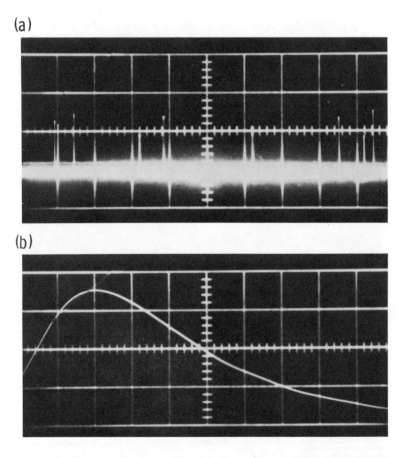

Figure 6-4. Oscilloscope traces of: (a) the pulses showing the burst of ions produced from the volatilizing and ionizing of individual dioctyl phthalate particles of 1 μm diameter; (b) the fast scan of a single pulse. The horizontal scale is 100 ms per division for (a) and 20 μs for (b). Mass peak at m/z 149 is monitored. (Reproduced with permission from ref. 14.)

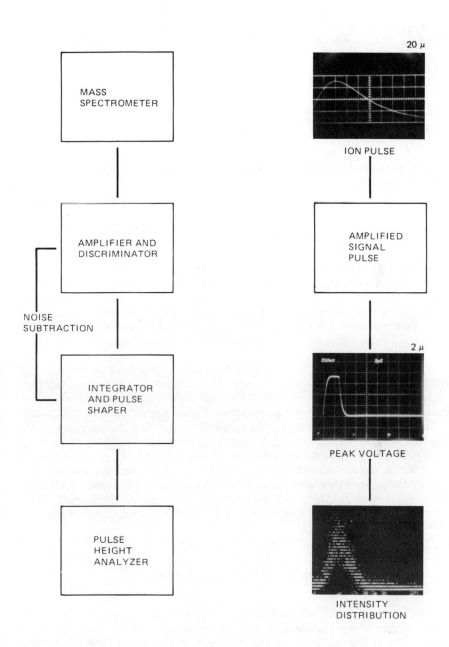

Figure 6-5. Average intensity measurement scheme for a mass peak.

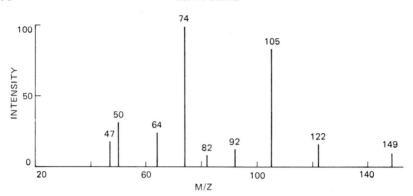

<u>Figure 6-6</u>. Mass spectrum obtained from potassium biphthalate particles of 1.7 μm diameter.

dilute solution of the compound through an orifice (10-20 μm diameter) in a plate; the resulting liquid jet is broken by vibrating the plate in contact with a piezoelectric crystal at a fixed frequency. This generates droplets of uniform size that, after drying in a stream of air or nitrogen, produce a highly monodispersed aerosol. The particle size can be changed by properly adjusting the concentration of the solution, its feed rate, and the frequency of vibration. A nebulizer was used for generating aerosols of bacteria particles, and for the resuspension of flyash a fluidized-bed aerosol generator was used.

Quantitative aspects of the PAMS technique has been investigated by using monodisperse aerosols. Beams of particles composed of different chemical species were made, and the intensities of their characteristic masses were measured for various size particles; Figure 6-7 summarizes the results. The mass spectrometer signal increases linearly with the increase in the particle volume, which indicates a complete vaporization of the particle in the V-type filament. Because this observation was based on the measurements of particles in a rather narrow size range (0.9-2.1 μm), the investigation needs to be extended to other particle sizes and to aerosols composed of different chemical species. However, it is reasonable, from the nature of the compounds studied to expect bacteria particles of 0.9-2.1 μm to be volatilized completely within the filament.

V. BACTERIAL PARTICLE ANALYSIS

In applying the PAMS technique to the analysis of biological particles, mass spectrometric measurements of different bacterial particles have been made. The common bacteria <u>Pseudomonas putida</u> and spores of <u>Bacillus subtilis</u> and <u>Bacillus cereus</u> have been analyzed. These studies demonstrate the

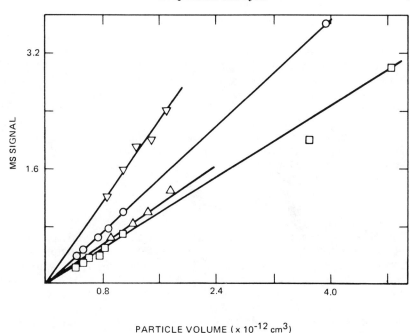

PARTICLE VOLUME ($\times 10^{-12}$ cm^3)

<u>Figure 6-7</u>. Variation of the mass spectrometer (MS) signal with the particle volume; ▽ denotes mass peak at m/z 86 from glutaric acid; O denotes mass peak at m/z 55 from adipic acid; □ denotes mass peak at m/z 80 from ammonium sulfate; △ denotes mass peak at m/z 149 from dioctyl phthalate. (Reproduced with permission from ref. 14.)

feasibility of analyzing individual bioparticles on a continuous, real-time basis. The preparation of aerosol samples, their introduction into the mass spectrometer, and the results obtained from them are summarized below.

A. Sample Preparation

A culture medium containing tryptone, yeast extract, glucose, and sodium chloride was used to grow the bacteria. The pH of the medium was adjusted initially to 7.8. <u>Pseudomonas putida</u> (ATCC 29607) bacteria were grown in liquid suspension at room temperature for 40 h under continuous agitation. Samples of <u>B. cereus</u> 203A and <u>B. subtillis</u> 168 were obtained from the UCLA Biology Department and were grown in liquid suspension for 48 h at 34°C. Sporulation was completed by placing the culture in a 70°C water bath for 30 min. The culture was harvested by repeated washing with water, and the water suspension of bacteria was refrigerated until needed.

It was observed that the aerosol generated from an aqueous medium generally resulted in a lower transmission of particles through the beam generator. The transmission efficiency was increased significantly if a volatile solvent (eg, alcohol) was used in place of water. The incomplete drying of water droplets containing particles seems responsible for the reduced transmission efficiency. In order to overcome this problem, bacterial suspensions in ethanol were prepared by washing the sample successively with a series of water-ethanol solutions of increasing ethanol concentration. The final suspension in pure ethanol was used for aerosolization. Larger aggregates of cells, formed during the washing by centrifugation, were removed by filtering the suspension through a 20-μm mesh screen.

B. Aerosolization and Beam Generation

The alcohol suspension of bacteria was aerosolized in a micronebulizer in a stream of nitrogen gas. Because the droplets acquired charge during their generation, they were passed through a cylinder containing Kr^{85}, where the charge was neutralized by capturing ions of gas molecules. The aerosol finally was dried by passing it through a silica gel diffusion drier (Figure 6-1). The particle concentration in the aerosol was about 70/mL determined with an optical particle counter. The electron micrographs in Figure 6-8 show that the bacterial aerosols are free from extraneous particles and debris and that the bacteria remain intact after aerosolization. The micrographs also indicate that the aerosol particles are predominantly single spores, 20% occurring as doublets.

After drying, the aerosol stream passes into the sampling chamber. A portion of the aerosol (50 atm mL min^{-1}) is isokinetically sampled into the beam tube, which, after expansion through the capillary nozzle, produces a beam of bacteria particles. The particles having higher inertia than the gas molecules form a tightly focused beam and are led into the mass spectrometer chamber through the skimmers, whereas most of the carrier gas is pumped off in this region. In order to characterize the physical state of the bacteria in the beam, the particles were collected on a glass slide in the mass spectrometer chamber. Particles in the beam attain high terminal velocity ($\sim$300 ms-1) [63, 64], which necessitates the coating of glass slide with Apiezon grease so that they do not bounce off the glass slide. A photograph of the composite of the particle deposit under low magnification is shown in Figure 6-9. The beam has a divergence of 1.6 x 10^{-3} sr. Light and electron micrographs of the particle beam deposit are also presented in Figure 6-10. It can be seen that Apiezon grease is not a suitable substrate for electron microscopy. However, the rod shape and size of the spores are clearly discernible, showing that the spores retain their integrity during the beam generation and particle collection processes. This is the first time that a beam of cells has been made in vacuum and directly introduced into the mass spectrometer. The generation of the beam is not limited to spores possessing hardy characteristics; P. putida is a nonspore-forming bacteria.

C. Mass Spectral Measurements

The bacteria particles in the beam strike the hot V-type rhenium filament (740°C) and are pyrolyzed into small mass fragments. A burst of ions is produced from individual particles after volatilization and ionization by electron impaction (~40 eV in energy) in the ion source of the mass spectrometer. The pulse width of 100-200 μs necessitates the monitoring of one mass peak from each particle that is manually set at the mass spectrometer control unit. The average intensity of a mass peak is obtained from the pulse-height distribution of about a thousand pulses from different particles. Signals were observed throughout the mass range of 30-300 atomic mass units. A set of 30 mass

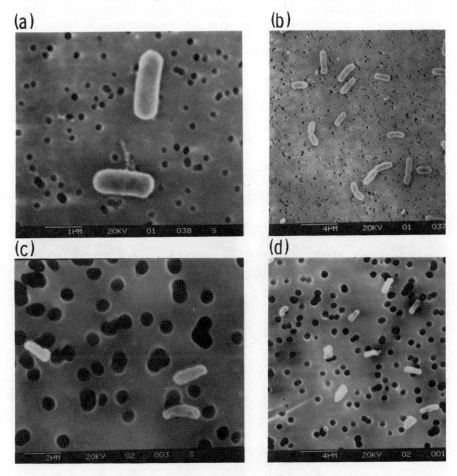

Figure 6-8. Electron micrographs of Bacillus subtilis; (a) and (b) show the cells taken directly from their suspension, whereas (c) and (d) are the cells collected from their aerosols after the diffusion drier.

peaks was selected and their intensities measured for all three species (Figure 6-11).

The intensities of the different mass peaks are normalized to the most intense peaks in their respective spectrum. A striking similarity between the spectra can be seen, which is not surprising because the microorganisms have essentially the same major chemical building blocks. However, some visual differences can be seen in them. The mass peaks between 192 and 275 atomic mass units were weak for P. putida and B. subtilis, and a good intensity measurement of these peaks could not be made. The spectra may be distinguished by considering the relative intensities of different peaks. For example, the relative intensities of mass peaks at m/z 67, 114, 169, and 282 compared to the intensity at m/z 149 are higher in the spectrum from B. cereus than those in the spectrum from P. putida. Another difference between the P. putida and the Bacillus samples lies in the presence of m/z 134 peak in the Bacillus samples.

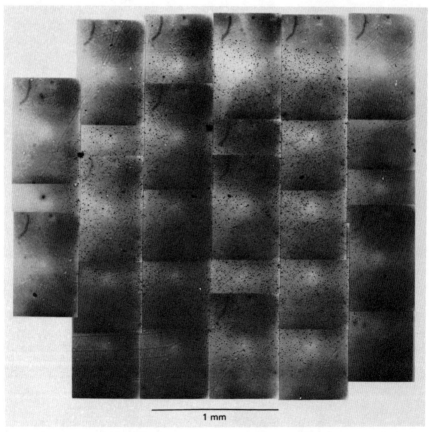

1 mm

Figure 6-9. Deposits of Bacillus subtilis particles from a beam on a glass slide coated with Apezion grease in the mass spectrometer chamber.

(a)

(b)

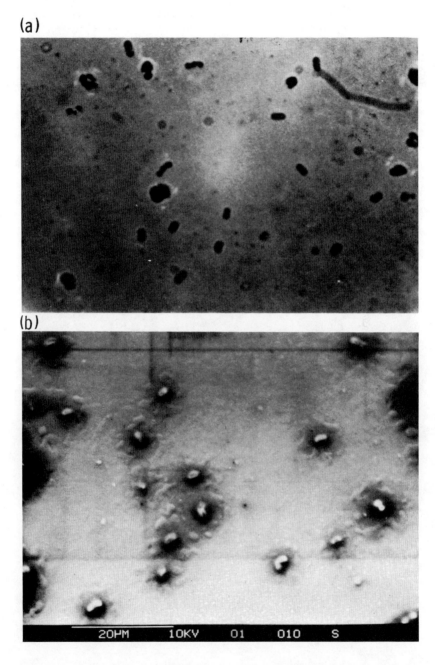

Figure 6-10. Microscope photograph and electron micrograph of <u>Bacillus sub-</u>
<u>tilis</u> from beams collected on Apezion grease-coated slides: (a) light microscope
photograph; (b) electron micrograph.

A more objective comparison of mass spectra and their reproducibility can be made by applying statistical procedures [75]. One such preliminary analysis of the mass spectra based on the criterion of the degree of correspondence [34] has been made. The correspondence between P. putida and the Bacillus samples is significantly low. A value of .64 for the degree of correspondence between P. putida and a reference spectrum of B. cereus shows that the two spectra can be differentiated easily. However, a variability of ~25% is calculated for the replicate samples of B. cereus. It is believed that the polydisperse nature of the aerosol and variability introduced by the signal

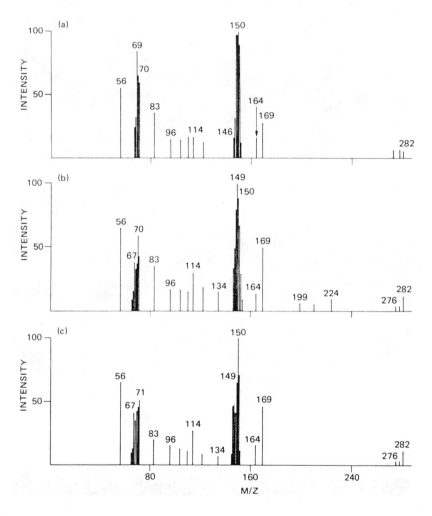

Figure 6-11. Mass spectra of bacteria particles: (a) Pseudomonas putida, (b) Bacillus cereus, (c) Bacillus subtilis. (Reproduced with permission from ref. 69.)

processing, as well as some variability in the samples themselves, may be the contributing factors. The details of the analysis are given by Platz [76].

D. Comparison with Other Pyrolysis Methods

The choice of P. putida, B. subtilis, and B. cereus for study with the PAMS technique was made mainly because PyMS data on them were available in the literature. Schulten et al [77] made measurements on P. putida. In their work, a continuous-mode pyrolyzer was used. A sample of 5 mg of freeze-dried bacteria was pyrolyzed in a flask for 2 min at 500°C and the pyrozylate was mass analyzed by field-ionization mass spectrometry. Field ionization, a softer method of ionization, produces predominantly molecular ions during ionization with minimal fragmentation. The spectrum so obtained, however, consists mainly of mass peaks of m/z less than 150. Boon et al [78] have made mass spectral measurements of B. subtilis. Their sample was volatilized in a Curie-point pyrolyzer. The pyrolyzate effused from the pyrolyzer through an orifice into the quadrupole mass spectrometer, where it was ionized with electrons of low energy (14 eV), and the spectrum was accumulated by repeated scanning of masses for approximately 10 s. Here also, the most intense peaks lay in the low-mass range (40-60) and were confined to m/z of less than 120. The fragmentation to low masses seems to result from the pyrolysis condition. Some fragmentation also may be possible during the collisions in the vapor phase. It must be added that the equilibrium temperature in the Curie-point pyrolyzer is attained very fast, which assures a reproducible uniform temperature pyrolysis condition. Very encouraging results in the detection and identification of microorganisms have been obtained by this method.

In contrast to the above results, abundant molecular fragments are produced at high masses in the PAMS methods even though the ionization is made by the impaction of electrons of higher energy (40 eV), and the transmission of ions through the quadrupole analyzer decreases with increasing mass. The larger mass fragments carry more information about their parents and consequently should help in analysis. It is believed that the primary reason for high mass peaks is the pyrolysis process. Each cell is volatilized separately, and many of the desirable conditions for pyrolysis stated in Section II.C are fulfilled. Individual particle pyrolysis is fast, with an estimated duration of less than 50 μs. A cell particle 0.5 μm in diameter and 1.5 μm in length is expected, after volatilization and fragmentation, to produce a pressure of less than 10^{-4} Pa in an ionization volume of 1 mL, so the vapor molecules are essentially in a collision-free condition in the ion source, eliminating any secondary reactions in the vapor phase. The pyrolyzing filament is situated within the ionizer and, therefore, minimizes the transit time between the pyrolyzer and the ion source. Mass peaks of various monodisperse organic and inorganic aerosols have been found to have a relative standard deviation of 10% in their intensities, showing the reproducibility of the pyrolysis method [14].

VI. CONCLUSIONS

Particle analysis by mass spectrometry has the following advantages:

1. Aerosol particles can be introduced directly into the ion source of a mass spectrometer in the form of a beam, thus eliminating the need of sample collection and preparation.

2. The bacteria particles remain intact during the process of beam generation.

3. Individual particles can be volatilized within a V-type filament.

4. The pyrolysis provides fast heating, uniform pyrolysis temperature, and proximity of the pyrolyzer to the ion source, and it assures that only one extremely small sample at a time is pyrolyzed in a collision-free environment.

5. Mass spectra of individual biological particles can be obtained on a continuous, real-time basis.

6. The mass spectra contain peaks at high masses, which can provide more information about their parents.

The results obtained on P. putida, B. subtilis, and B. cereus demonstrate the potential of the PAMS technique for the detection and identification of biological particles. The preliminary nature of the work should be emphasized, however. More work on different biological samples needs to be done. The advanced data analysis technique of chemical pattern recognition developed by various workers [42] should be incorporated in the technique.

It is encouraging to note that other pyrolysis-mass spectrometric methods for analyzing biological material have found considerable success. In view of this and the advantages of the PAMS technique, it is safe to conclude that the PAMS method holds promise for biological particle analysis. The real-time analytical capability of the method makes it uniquely suitable for monitoring a protected environment (eg, a clean room) or a general field environment.

VII. FUTURE DEVELOPMENTS

One of the limitations of the PAMS method arises from the mode of its data acquisition. One mass peak is measured per particle. In order to obtain the average intensity of a mass peak from a polydisperse aerosol, a large number of particles must be measured, and the process must be repeated for all the mass peaks. This entails a long analysis time, particularly for the ambient air sample, where the number density of biological particles is expected to be low. This limitation is inherent not in the PAMS technique but in the use of a quadrupole mass spectrometer. A quadrupole mass spectrometer measures the intensities of different masses by scanning them in time and is too slow for a complete scan of an ion pulse from a particle. Moreover, a scanning instrument is, in general, less sensitive than a nonscanning one because in the scanning mode, a particular signal is measured for only a small part of its total time of generation. However, in a restricted environment where not much interference exists, one characteristic mass peak, possibly in the high-m/z range, could be selected for the particle, and a quadrupole instrument could still be used for the detection. The problem associated with the use of a quadrupole mass spectrometer could be overcome by replacing it with a focal-plane mass spectrograph (nonscanning); then, all the ions of different masses arising from a single particle could be measured simultaneously.

A miniaturized mass spectrograph (Mattauch-Herzog type) with an electronic detector, known as electrooptical ion detector [79], covering the entire focal plane has been developed and built at the Jet Propulsion Laboratory. Figure 6-12 shows a photograph of this instrument. It has a 5-in.-long focal plane and covers a mass range of 28-500 mass units. The mass spectrograph with the particle beam generator is being assembled. This system will provide a complete mass spectrum from individual particles. A computerized pattern-recognition technique of analysis will help differentiate the mass spectra of different microorganisms.

Because lasers have been found to provide a controllable and efficient method of ionization [80], they may help generate characteristic mass spectra (fingerprints) of different particles. Preliminary results on the application of laser-induced volatilization and ionization of individual particles in the beam on the PAMS system have been obtained. A beam of potassium biphthalate particles was generated, and individual particles in the beam were hit by high-energy Nd-YAG laser pulses for their volatilization and ionization [64]. Mass spectral measurements show the production of ions at masses 39, 28, and 23. The signal at m/z 39 originates from K^+, and its combined efficiency is found to be about 10^{-6}. Signals at mass 28 and 23 were weaker in intensity. The former is believed to arise because of mass fragments CO from the carboxylic groups and the latter from sodium present as an impurity. The lower ionization efficiency and the absence of other mass signals from the organic part of KBP molecules are attributed to the low laser power density ($\sim 2 \times 10^5$ W/cm^2) as well as to the incomplete vaporization of the particle. Work is continuing in the study of laser-induced ionization with increased power density on the particles while in the beam.

A library of mass spectral data of the different microorganisms should be compiled for identification by comparison. Biological particles in air are expected to be mixed with nutrients, attached to dust particles, or mixed with other substances that obscure their identity; sometimes more than one kind of particle may be mixed together. Such interferences also should be studied.

Figure 6-12. Photograph of a miniaturized mass spectrograph with the electrooptical ion detector. (Courtesy of C.E. Giffin.)

ACKNOWLEDGMENTS

The work on particle analysis by mass spectrometry was sponsored by the National Science Foundation and the U.S. Army Research Office. This has been a collaborative effort between the Jet Propulsion Laboratory, California Institute of Technology, and the University of California at Los Angeles.

The author wishes to thank Ms. Rose Carden for typing this chapter.

VIII. REFERENCES

1. Artenstein, M.S.; Lamson, T.H. In "Aerobiology", Proceedings of the Third International Symposium, 1969; Silver, I.H., ed.; Academic Press: London, 1970; p. 1.
2. Fraser, D.W.; Tsai, T.F.; Orenstein, W.; Parkin, W.E.; Beecham, H.J.; Sharrara, R.G.; Harris, J.; Mallison, G.F.; Martin, S.M.; McDade, J.E.; Shepard, C.C.; Bracham, P.S. N. Engl. J. Med., 1977, 297, 1189.
3. Muir, D.C.F. In "Clinical Aspects of Inhaled Particles"; Muir, D.C.F., ed.; Heinemann: London, 1972; p. 1.
4. Gregory, P.H. "The Microbiology of the Atmosphere"; Interscience Publishers: New York, 1961; p. 33.
5. Cox, C.S. J. Gen. Microbiol., 1966, 43, 383; 1966, 45, 283.
6. Cinkotai, F.F.; Lockwood, M.G.; Rylander, R. Am. Ind. Hyg. Assoc. J., 1977, 38, 554.
7. Kleyn, J.G.; Johnson, W.M.; Wetzler, T.F. Appl. Environ. Microbiol., 1981, 41, 1454.
8. Dart, R.K.; Stretton, R.J. "Microbiological Aspects of Pollution Control"; Elsevier Scientific Publishing Co.: Amsterdam, 1980; Chap. 2, pp. 42-53.
9. Johnson, D.E.; Caman, D.E.; Sorber, C.A.; Sagik, B.P.; Glennon, J. P. In "Proceedings of Risk Assessment and Health Effects of Land Application of Municipal Wastewater and Sludges", Sagik, B.O.; Sorber, C.A., eds.; Center for Applied Research and Technology: San Antonio, TX, 1978, pp. 240-271.
10. Lighthart, B.; Frisch, A.S. Appl. Environ. Microbiol., 1976, 31, 700.
11. Fincher, E.L. In "An Introduction to Experimental Aerobiology"; Dimmick, R.L.; Akers, A.B., eds.; Wiley-Interscience: New York, 1969; Chap. 17, pp. 407-436.
12. Tyrrell, D.A.J. In "Airborne Microbes", Seventeenth Symposium of the Society of General Microbiology; Gregory, P.H.; Monteith, J.L.; eds.; Cambridge University Press: Cambridge, 1967; p. 247.
13. Burke, J.F.; Corrigan, E.A. New Engl. J. Med., 1961, 264, 231.
14. Sinha, M.P.; Giffin, C.E.; Norris, D.D.; Estes, T.J.; Vilker, V.L.; Friedlander, S.K. J. Colloid Interface Sci., 1982, 87, 140.
15. Greene, V.W. Environ. Sci. Technol., 1968, 2, 104.
16. Barnett, M.I. Ann. N.Y. Acad. Sci., 1969, 158, 674.
17. Sall, T. Trans. N.Y. Acad. Sci., 1964, 26, 177.

18. Levin, G.V.; Heim, A.H. In "Life Sciences and Space Research", Vol. 3; Florkin, M. ed; North Holland Publishing Co., Amsterdam, 1965; pp. 105-119.
19. Mitz, M.A. Ann. N. Y. Acad. Sci., 1969, 158, 651.
20. Gebhardt, L.P. "Microbiology"; The C.V. Mosby Company: St. Louis, 1970; Chap. 10, pp. 109-123.
21. Cherry, W.B.; Goldman, M.; Carski, T.; Moody, M.D. "Fluorescent Antibody Technique in the Diagnosis of Communicable Diseases"; U.S. Public Health Service Publication No. 729, 1960.
22. Beautner, E.H. Ann. N. Y. Acad. Sci, 1971, 177, 1.
23. Cherry, W.B.; Moody, M.D. Bacteriol. Rev., 1965, 29, 222.
24. Walwick, E.R.; Kay, R.E.; Zalite, B.R. In "Life Sciences and Space Research" Vol. 5; Brown, A.H.; Favorite, F.G. eds.; North Holland Publishing Co., Amsterdam, 1967; pp. 229-238.
25. Rudkin, G.T. In "Submicrogram Experimentation"; Cheronis, N.D., ed. Interscience: New York, 1961.
26. Bovallius, A.; Bucht, B.; Casperson, T.; Lundin, T.; Ritzen, M. Forsvarsmedicin, 1968, 4, 85.
27. Holmsen, H.; Holmsen, I.; Bernhardsen, A. Anal. Biochem., 1966, 17, 456.
28. Sharpe, A.N.; Woodrow, M.N.; Jackson, A.K. J. Appl. Bacteriol., 1970, 33, 758.
29. Johnson, R.A.; Hardman, J.G.; Broadus, A.E.; Sutherland, E.W. Anal. Biochem., 1970, 35, 91.
30. Bruch, C.W. In "Airborne Microbes"; Gregory, P.H.; Monteith, J.L., eds.; Cambridge University Press: Cambridge, 1967; pp. 345-374.
31. Henis, Y. et al. Appl. Microbiol., 1966, 14, 513.
32. Mitruka, B.M.; Alexander, M. Ann. Rev. of Biochem., 1967, 20, 548.
33. Meuzelaar, H.L.C.; Kistemaker, P.G.; Tom, A. In "New Approaches to the Identification of Microorganisms"; Heden, C.G.; Illeni, T., eds.; John Wiley and Sons: London, 1975; Chap. 10, pp. 165-177.
34. Kistemaker, P.G.; Meuzelaar, H.L.C.; Posthumus, M.A. In "New Approaches to the Identification of Microorganisms"; Heden, C.G.; Illeni, T., eds.; John Wiley and Sons: London, 1975; Chap. 11, pp. 179-191.
35. Simmonds, P.G. J. Appl. Microbiol., 1970, 20, 567.
36. Cummins, C.S.; Harris, H. J. Gen. Microbiol., 1956, 14, 583.
37. Hill, L.R. In "Identification Methods for Microbiologists"; Gibbs, B.M.; Shapton, D.A., eds.; Technical Series No. 2, B, The Society for Applied Bacteriology: London, 1968, pp. 177-186.
38. Irwin, W.J. J. Anal. Appl. Pyrol., 1979, 1, 30; 1981, 3, 3.
39. Morgan, S.L.; Fox, A. In "Chemotaxonomic Characterization of Microorganisms and Chemical Detection of Infectious Diseases by Capillary GC, Pyrolysis GC-MS and Solid Phase RIA", 2nd ARO Biodetection Workshop, North Carolina State University, Raleigh, N.C.; July 13-15, 1982; Sponsored by U.S. Army Research Office, N.C.; and references therein.
40. Gonser, G.L.; Heck, H.D.A.; Arbar, M. Anal Biochem., 1976, 71, 519.
41. Schulten, H.R. In "New Approaches to the Identifications of Microorganisms"; Heden, C.G.; Illeni, T., eds.; John Wiley and Sons: London, 1975;

Chap. 9, pp. 155-164.
42. Wieten, G.; Meuzelaar, H.L.C.; Haverkamp, J. In "Advances in Gas Chromatography/Mass Spectrometry"; Odham, G.; Larson, L. and Mardh, P.A., eds.; Plenum Publishing Co.: London, 1983; Chap. 9.
43. Meuzelaar, H.L.C.; Ficke, H.G.; Den Harmik, H.C. J. Chromatogr. Sci., 1975, 13, 12.
44. Meuzelaar, H.L.C.; Kistemaker, P.G.; Eshuis, W.; Engle, H.W.B. In "Rapid Methods and Automation in Microbiology"; Newson, S.W.B.; Johnston, H.H., eds.; Learned Information: Oxford, 1976; pp. 225-230.
45. Heinen, H.J. Int. J. Mass Spectrom. Ion Phys., 1981, 38, 309.
46. Denoyer, E.; Van Grieken, R.; Adams, F.; Natusch, D.F.S. Anal Chem., 1982, 54, 26A.
47. Hercules, D.M.; Day, R.J.; Balasanmugam; Dang, T.A.; Li, C.P. Anal. Chem., 1982, 54, 280A.
48. Hillenkamp, F.; Kaufmann, R.; Nitsche, R.; Remy, E.; Unsold, E. In "Microprobe Analysis as Applied to Cells and Tissues"; Hall, T.; Echlin, P.; Kaufmann, R., eds.; Academic Press: London, 1974; pp. 1-14.
49. Kaufmann, R.; Wieser, P. "Laser Microprobe Mass Analysis (LAMMA) in Particle Analysis"; paper presented at the 13th Annual Conference on the Microbeam Analysis Society, Ann Arbor, MI, June 22, 1978; published by U.S. Natl. Bur. Stand. April 1980; In "Characterization of Particles" Heinrich, K.F.J., ed.; Special Publication 533, pp. 199-233.
50. Van Der Peyl, G.J.Q.; Haverkamp, J.; Kistemaker, P.G. Int. J. Mass Spectrom. Ion Phys., 1982, 42, 125.
51. Cotter, R.J.; Tabet, J.C. Int. J. Mass Spectrom. Ion Phys., 1983, 53, 151.
52. Reiner, E. Nature (London), 1965, 206, 1271.
53. Menger, F.M.; Epstein, G.A.; Goldberg, D.A.; Reiner, E. Anal. Chem., 1972, 44, 424.
54. Coupe, N.B.; Jones, C.E.R.; Perry, S.G. J. Chromatogr., 1970, 47, 291.
55. Zemany, P.D. Anal. Chem., 1952, 24, 1709.
56. Meuzelaar, H.L.C.; Kistemaker, P.G. Anal Chem., 1973, 45, 587.
57. Meuzelaar, H.L.C.; Haverkamp, J.; Heileman, F.D. "Pyrolysis Mass Spectrometry of Recent and Fossil Biomaterials"; Elsevier Scientific Publishing Co.: Amersterdam, 1982.
58. Akers, A.B.; Won, W.D. In "An Introduction to Experimental Aerobiology"; Dimmick, R.L.; Akers, A.B., eds.; Wiley-Interscience: New York, 1969; Chap. 4, pp. 59-99.
59. Myers, R.L.; Fite, W.L. Environ. Sci. Technol., 1975, 9, 334.
60. Davis, W.D. Environ. Sci. Technol., 1977, 11, 587.
61. Chatfield, D.A. In "Proceedings of 30th Annual ASMS Conference on Mass Spectrometry and Allied Topics"; Honolulu, Hawaii, 1982, pp. 60-61.
62. Estes, T. J.; Vilker, V.L.; Friedlander, S.K. J. Colloid Interface Sci., 1983, 93, 84.
63. Sinha, M.P.; Chatfield, D.A.; Platz, R.M.; Vilker, V.L.; Friedlander, S.K. In "Proceedings of 30th Annual ASMS Conference on Mass Spectrometry and Allied Topics", Honolulu, 1982, pp. 352-53.
64. Sinha, M.P. Rev. Sci. Instrum., 1984, 55, 886.

65. Murphy, W.K.; Sears, G.W. J. Appl. Phys., 1964, 85, 1986.
66. Israel, G.W.; Friedlander, S.K. J. Colloid Interface Sci., 1967, 24, 330.
67. Dahneke, B.E.; Friedlander, S.K. J. Aerosol Sci., 1970, 1, 325.
68. Dahneke, B.E. In "Recent Developments in Aerosol Science"; Shaw, D.T., ed.; John Wiley and Sons: New York, 1978.
69. Sinha, M.P.; Platz, R.M.; Vilker, V.L.; Friedlander, S.K. Int. J. Mass Spectrom. Ion Processes, 1984, 57, 125.
70. Friedlander, S.K. "Smoke, Dust and Haze: Fundamentals of Aerosol Behavior"; John Wiley and Sons: New York, 1977.
71. Taulbee, D.B. J. Aerosol Sci., 1978, 9, 17.
72. Johnson, C.E.; Sinha, M.P.; Friedlander, S.K. In "Feasibility Studies of the Use of PAMS for Flyash Particles"; Electric Power Research Institute Contract No. TPS81-783, 1983; Final Report.
73. Sinha, M.P.; Platz, R.M. Aerosol Sci. Technol., 1983, 2, 256; annual conference issue of American Association for Aerosol Research.
74. Berglund, R.N.; Liu, B.Y.H. Environ. Sci. Technol., 1973, 7, 147.
75. Windig, W.; Kistemaker, P.G.; Haverkamp, J. J. Anal. Appl. Pyrol., 1981, 3, 199.
76. Platz, R.M. In "Particle Analysis by Mass Spectrometry for Detection of Single Bacteria in Air Suspension"; M.S. Thesis, Department of Chemical Engineering, University of California at Los Angeles, 1983.
77. Schulten, H.R.; Beckey, H.D.; Meuzelaar, H.L.C.; Boerboom, A.J.H. Anal. Chem., 1973, 45, 191.
78. Boon, J.J.; DeBoer, W.R.; Kruyssen, F.J.; Wouters, J.T.M. J. Gen. Microbiol., 1981, 12, 119.
79. Boettger, H.G.; Giffin, C.E.; Norris, D.D. In "Multichannel Image Detectors"; Talmi, Y., ed.; ACS Symposium Series No. 102, American Chemical Society: Washington, D.C., 1976; pp. 292-317.
80. Conzemius, R.J.; Cappelen, J.M. Int. J. Mass Spectrom. Ion Phys., 1980, 34, 197.

7. DETECTION OF MICROORGANISMS AND THEIR METABOLISM

BY MEASUREMENTS OF ELECTRICAL IMPEDANCE

W. Keith Hadley and David M. Yajko

I. INTRODUCTION

A decrease in the electrical impedance of a culture occurs when actively metabolizing bacteria utilize large molecules in a culture medium and form smaller ion pairs. The sensitive instruments available now [1-7] usually detect actively metabolizing bacteria when 10^6 to 10^7 bacteria per millileter are present in the culture. These instruments have been used for detection of bacteria or yeasts in clinical specimens, for food analyses, for industrial microbial process control, and for sanitation microbiology. By standardizing the instrument, culture medium, and sample it is possible to determine the approximate number of microorganisms present in the original sample by the impedance detection time. This is the time elapsing between inoculation of the sample into the monitored culture module and the detection of a change in impedance. By measuring in turn the impedance of enrichment cultures, selective media, and media containing specific substrates and antimicrobials, it is possible to detect the presence of certain kinds of bacteria, identify bacteria, and determine their antimicrobial susceptibility. Conductimetry, a similar electrical measurement, has been used for measurement of enzymatic activity [8,9]. Commercial conductimetric instruments are being developed [10].

This chapter considers the principles of electrical impedance, the history of electrical measurements of bacterial cultures, the advantages and disadvantages of electrical impedance measurement, and the different instruments available for microbial analysis. The applications of impedimetry are examined and analyzed. Future uses of electrical impedance analysis are considered.

A. Electrical Impedance

Impedance is a measure of the total opposition to the flow of a sinusoidal alternating current in a circuit containing resistance, inductance, and capacitance. Inductance and capacitance together are called the reactive part of the circuit. The changes in impedance that occur in a microbial culture can be measured by placing two metal electrodes into the culture medium and introducing an alternating current into the circuit. Impedance usually is measured by a bridge circuit. Often a reference module is included to

193

measure and exclude non specific changes in the test module. The reference often contains uninoculated culture medium. The reference module serves to control for temperature changes, evaporation, changes in amounts of dissolved gases, and degradation of culture medium during incubation. The impedance of such a culture can be represented by a combination in series of the impedance of the culture and the polarization impedance at the interface of the culture with the electrodes. Between 400 HZ and 25 KHZ, culture medium impedance is mainly resistive (the reciprocal of the conductance). The instruments available for analyzing microbial cultures usually measure not complex impedance but some function of it. Any change in the components of a circuit, including bacteria or their metabolic products, changes the impedance and alters the voltage-current relationship. Differences in the circuit become more marked at lower frequency. These variations in impedance occur when bacteria, plant, or animal cells are in suspension in the circuit [11 -15].

The relationship between voltage (E) and current (I) for a given frequency (f) when a circuit contains both a resistor (R) and a capacitance (C) is given by:

$$E = I(R^2 + X_c^2)^{1/2} \qquad (7\text{-}1)$$

where X_c is the capacitative reactance and is given by:

$$X_c = 1/2\,\pi f C \qquad (7\text{-}2)$$

Stacey [16] provides further discussion of these relationships.

A model of a resistor and a capacitor in series serves as a first approximation to describe the electrical circuit formed when a pair of metal electrodes is placed into a microbiological medium. The relationship between impedance (Z), resistance (R), capacitance (C), and frequency (f) for a resistor and a capacitor in series is expressed as follows:

$$Z^2 = R^2 + 1/(2\,\pi f C)^2 \qquad (7\text{-}3)$$

B. History of Electrical Measurements of Microbial Metabolism and Growth

Only a scattering of early studies used electrical measurements of microbial cultures to study metabolic activity and growth. These studies were all done with alternating current, because direct current would cause electrolysis in the culture. Measurements of impedance were made by methods that were very insensitive and the data were expressed in conductivity terms.

Stewart described the increasing electrical conductivity of putrifying defibrinated blood to the British Medical Association at Edinburgh in 1898 and published the work the following year in the Journal of Experimental Medicine [17]. Oker-Blom [18] in 1912, in similar experiments, found a tenfold change in conductivity of putrifying blood over a 25-d period.

Parsons and co-workers, working for a meat packing company, used electrical conductivity measurements to follow the metabolic activity of clostridia incubated anaerobically in various broths. These studies were reported from 1926 to 1929 [19-21]. They were able to correlate conductivity measurements with ammonia production. Sierakowski and Leczycka, in 1933, concluded that much of the increased conductivity and ammonia production observed after many days of incubation was associated with cell lysis [22].

In 1938, Allison and associates showed that the extent of bacterial proteolysis was closely related to conductimetric measurements [23]. Their apparatus could measure impedance to 0.1 ohm. With this increased sensitivity, significant changes were detectable at shorter intervals of time.

Twenty years later, McPhillips and Snow studied the relationship between acid production in milk by Streptococcus lactis and culture conductivity using a unique torroidal conductivity cell [24]. The torroidal cell had no electrodes dipping into the milk culture and thus avoided electrode polarization impedance. Conductivity was calculated from the changes of electrical properties of currents flowing through coils wrapped around opposite ends of the torus. Measurements were made hourly and resulting curves of conductivity vs time were similar to the impedance vs time plots produced with the more sensitive modern impedimetric instruments.

II. SYSTEMS FOR MICROBIAL ANALYSIS BY ELECTRICAL IMPEDANCE AND CONDUCTANCE

A listing of modern instruments and their sources appears in Table 7-1.

A. The Bactobridge

Ur first described the Bactobridge (Table 7-1) in studies on blood coagulation [25-27]. The application to microbiology was recognized in the first article. Subsequent studies by Ur and Brown described the application to microbiology [1,28,29]. The principle of the electrical-impedance-measuring instrument is simple. Comparative impedance is measured in two conductivity cells on a bridge circuit. One cell contains the sample to be measured, eg, clotting blood or a bacterial culture. The balancing resistance consists of a cell containing a similar material in which the process to be measured is not occurring, eg, heparinized blood from the same person or uninoculated bacterial culture medium. All impedance changes are therefore balanced, but

<u>Table 7-1.</u> Manufacturers and Developers of Instruments

---

1. Bactobridge
   TEM, Centronic Sales
   King Henry's Drive
   New Addington, Croydon CR 9 OBE
   England, U.K.

   Manufacturers and suppliers of Bactobridge instruments

2. Bactomatic
   A Division of Medical Technology Corporation
   719 Alexander Road
   P.O. Box 3103
   Princeton, New Jersey 08540, USA

   Bactomatic Ltd.
   New Town Industrial Estate
   Newtown Road
   Henley-on-Thames
   Oxon, RG9 1HG England, U.K.

   Manufacturers and suppliers of Bactometer instruments

3. Honeywell Inc.
   Gerald J. Wade Mail #108
   4800 E. Dry Creek Road
   Littleton, Colorado 80122, USA

   Patents held on electrode configurations and detection systems.
   Honeywell does not manufacture or sell instruments for impedance
   measurements of bacterial cultures.

4. Japan Tectron Instrument Corporation

   Manufacturers and suppliers of the Orga 6 Impedance Measuring
   Instrument

5. Malthus System
   Malthus Instruments Ltd., Site 1
   Almondbank, Perthshire, PH1 3NQ, U.K.

   Manufacturers and suppliers of the Malthus System

---

for those resulting from coagulation or bacterial metabolism or growth. The
Bactobridge uses a carefully matched pair of special conductivity cells, which
have resistivity within 10 ohms of each other and well-matched capacitance
and thermal properties. Each cell has a 100 μL volume and gold-plated
electrodes and contacts. The bridge is activiated by a 10-KHz alternating

current. A pure sinusoidal signal and maintenance of voltage below 0.5 v peak to peak reduces polarization effects. All components are kept in an incubator chamber. Results are recorded on a chart recorder or computer. Ur and Brown were able to detect an initial bacterial inoculum of 10^5 to 10^3 colony-forming units per milliliter within 1-3 h incubation. Media with less buffering capacity, eg, PPLO broth, showed impedance change more rapidly than highly buffered media, such as tryptose phosphate broth. The susceptibility of bacteria to antimicrobials was detectable within 3 h. Less time was required for a bacterial culture to accumulate metabolites in the small 10-µL cell used for the Bacto-bridge than in a larger cell. However, the small size may be a disadvantage for inocula that contain very small numbers of bacteria, eg, blood.

B. The Bactometer

Bactometer is the registered trademark for a series of automated systems for microbiological analysis that are made by Bactomatic (now a division of Medical Technology Corporation). The original U.S. patents 3,743,581 (1973), 3,890,201 (1975), 4,067,951 (1978), were issued to Paxton Cady and William J. Welch and assigned to Bactomatic. These instruments and modifications have been extensively described by Cady and others [2-5], and have been used in clinical studies for detection of microbial growth in blood, cerebrospinal fluid, and urine, and for antimicrobial susceptibility testing. The instrumental system has been used extensively for food, sanitation, and industrial process control.

The instruments make measurements using a test chamber and an uninoculated reference chamber. The same medium is used in both chambers. In the Bactometer models 8 and 32, each test chamber is paired with a reference chamber and the instruments measure changes in the impedance ratio. The Bactometer model 120 uses a single reference chamber for each 15 test chambers. A microprocessor compensates for nonmicrobial impedance changes.

The system consists of an oscillator, an amplifier, a comparator circuit, a digitizing circuit and address system, and an incubator component with female connectors to accept the test and reference chambers. Various test chamber modules with circuits and electrodes have been used with the Bactometer. These vary from cylindrical chambers which hold 2.7 mL and have vertical electrodes extending from the bottom of the chamber to 100 mL culture bottles with a rubber septum cap and electrodes which extend through the septum towards the bottom of the bottle. Some of these modules are supplied as disposable, sterile units.

The Bactometer M-123 can retain and measure 128 samples. It incorporates an electronic analyzer-incubator, a microcomputer video display with color-coded representation of data, a dual disk drive, a printer, and test card reports.

C. Honeywell Inc.: Gerald J. Wade Patents

Honeywell Inc. supported the investigation of microbiological assay by electrical impedance measurement for several years. Gerald J. Wade was the scientific leader of this project. Two main patents resulted: the automated culture growth and detection system [30] and the medical specimen culture bottle [31]. Solutions were found to some of the problems of electrical impedance measurement of bacterial cultures. Agitation of the culture was used. This speeds the growth rate of bacteria and diminishes the time required for them to grow to the threshold at which the instrument can detect their presence. Mixing of the culture also disperses bacteria that may grow in colonial arrangements away from the electrodes and so escape early detection. Most other microbial culture impedimetry systems have no intrinsic shaking mechanism and measurements are usually made on stable cultures. Erratic measurements occur when air, other culture atmospheres, or gases evolved by cultures mix over the solid-phase surface of the electrodes of these instruments. Resumption of stability of measurement may require 30-60 min. The Honeywell-Wade system avoids this "trijunction effect" by total immersion of the electrodes in the culture and the separation of the culture head gas from the electrodes. The electrical measurement depends on charge dispersion through the double layer on the electrode.

D. Japan Tecktron Instruments Corporation

The Orga 6 Automatic Microbial Analyzer consists of an incubator for culture modules on which repetitive electrical impedance measurements can be made, the power supply and impedance monitor, and a recorder. Seventy-two culture tests can be done simultaneously, each with its uninoculated medium control. The instrument is designed for use as a rapid (3 h.) means for testing the antimicrobial susceptibility of microorganisms.

E. Goldschmidt and Wheeler System: University of Texas

An instrument system that uses electrical impedance measurement for detecting bacteriuria [32] yet is different from the other systems described here has been developed by Goldschmidt and Wheeler at the University of Texas, Houston. This system has not been marketed. Bacteria in urine have been detected by this analytical system without the intervening culture period that is required by conventional clinical microbiology procedures and the other impedance measuring systems described above.

The system uses a tetrapolar electrode and measures the effects of washed bacterial cells suspended in distilled water rather than their metabolites. Only one reading is taken, which correlates directly with the number of bacteria in the specimen. The system includes a square waveform generator, a voltmeter, a 10-KHz bandpass, an oscilloscope, a four-pronged probe, and stainless steel or platinum electrodes fastened to the cover of a petri dish. A correlation was found between the concentration of bacteria and the output voltage on calibration suspensions of Escherichia coli and patient specimens.

F. Malthus System

The Malthus system was described in 1978 by Richards and associates [10]. The Malthus instruments monitor conductance rather than impedance. The prototype instrument uses 10.0-mL glass tubes with screw caps. The molded caps have platinum-screened, ceramic, reusable electrodes that dip into the medium in the tube and connect to the Malthus meter. The tubes are filled with 9.0 mL of broth medium prewarmed to 37°C in a water bath, and the specimen is added. The tubes are scanned by the meter at short time intervals and the conductance measurements are recorded in a microcomputer. An accelerating change in conductance of 10 μS is recorded as the detection of microorganisms. Detection time then is related to the log viable count of bacteria per milliliter.

G. Electrochemical Detection of Bacteria with Platinum Electrodes

Wilkins has described an electrochemical method for detecting bacteria based on the time of evolution of hydrogen [33]. This is not an impedance measurement system. However, impedance measurement may be affected by factors similar to those which operate in this system and investigators of impedance should be aware of this work. The system consists of a platinum and reference electrode or two platinum electrodes with surface area ratios of four to one connected to a strip chart recorder [34,35]. Hydrogen evolution was measured by an increase in voltage in the negative (cathodic) direction. Linear relationships were established between the inoculum size and detection time for some bacteria. Hydrogen production can be detected by this system with Escherichia, Citrobacter, Enterobacter, Klebsiella, Proteus and other hydrogen-producing bacteria. The strip-chart record shows a lag period followed by a sharp increase in voltage to peaks of 100 to 250 mV and then a decline. Gram-positive bacteria such as Staphylococcus aureus and Staphylococcus epidermidis, Streptococcus pyogenes and Streptococcus faecalis as well as the Gram-negative bacteria Serratia marcescens and Pseudomonas aeruginosa had a response pattern with a lag phase followed by a gradual increase to a maximum 50 mV. The principal behind detection of these bacteria has not been established.

III. APPLICATIONS

A. Impedance Systems for Rapid Automated Diagnosis of Bacteremia

Bacteremia is rapidly life threatening, yet, with timely and appropriate antimicrobial therapy it often can be controlled rapidly. For this reason, the blood culture is one of the more important tests done by clinical microbiology laboratories. In conventional processing of blood cultures, much time is spent making smears and subcultures from culture bottles, ie, performing tests that will eventually be proved negative. The usual yield of positive cultures is 10-15% of the total sent to the laboratory. A screening system for blood cultures would save much of a technologist's time and allow the concentration of effort on cultures likely to be positive. The system should yield the minimum number of false negatives. Impedimetric systems that offer continuous monitoring of a culture bottle through indwelling electrodes are well adapted to this type of screening, because reentry and sampling introduces the risk of contamination of the culture.

Impedimetric systems have been studied for blood culture screening. Spector and associates [36] studied mock blood cultures set up with human blood and approximated the numbers of bacteria likely to be present in blood during bacteremia. They compared the results from the Bactometer 32 impedance system with those from the Bactec radiometric $^{14}CO_2$ detection system. Similar detection times for positive cultures were found with these two systems. Carbon dioxide gassing of cultures containing staphylococci and streptococci diminished the detection time for cultures containing these bacteria. The continuous monitoring provided by the Bactometer was judged to be an advantage over the batch monitoring at 1- to 4-h intervals done with Bactec. However, the 32 channels of the Bactometer were insufficient for continuous monitoring of the large number of blood cultures most clinical microbiology laboratories must process.

Buckland and co-workers [37] also used simulated blood cultures set up in the 2.6 mL modules provided with the Bactometer 32 system. Using rather large inocula, 100 colony forming units (CFU) per millileter, approximately 12 h were required for detection by change in impedance. No false positives or negatives were observed.

Hadley and associates [3,38] developed a blood culture screening system in association with Bactomatic using the Bactometer 32 system. A clinical study of this system was done on a total of 1801 blood cultures from patients at San Francisco General Hospital, run in parallel by the hospital's conventional system and by Bactometer 32 impedimetry. One bottle was processed in the conventional manner with aerobic and anaerobic subcultures, and Gram-stained smears at 24 and 72 h and 7 d, whereas the other bottle was not shaken, vented, or opened during the 7 d it was subjected to frequent impedance ratio monitoring with a Bactometer 32 instrument. Each bottle contained 90 mL of supplemented tryptic digest of casein-soy base medium and 10 mL of blood. A 20-mL sample of blood was drawn from each adult patient when a blood culture was useful for clinical diagnosis and divided equally into the two bottles. The Bactometer bottle contained two stainless steel wire electrodes that extended from the top of the bottle into the medium. Of the 1801 cultures examined 17.2% were positive by at least one method and 1.1% of the total Bactometer cultures were false negatives. That is, there was no Bactometer impedance detection, but a bacterium was subcultured from the bottle after the 7 day monitoring. Some of these were significant isolates. Most frequently the organism responsible for the false negatives grew at the bottom of the bottle or in a colonial growth away from the electrode. When the bottles were shaken, some of these were detected. Pseudomonas aeruginosa and Candida parapsilosis may not have grown to detection concentration in unvented cultures. By Bactometer 32 impedance monitoring, twenty-six (1.4%) total cultures were false positives. Some of these may have been true positives, because mycoplasma cultures were not done. The Bactometer 32 blood culture system had a sensitivity of 93.6% and specificity of 98.0%. All bacteria except enterococci were detected more rapidly by the Bactometer than by the conventional system. All Bactometer cultures that showed an impedance decrease (except the false positives) had detectable bacteria on Gram-stained smears. The Bactometer 32 blood culture system was an effective screening system for detecting positive cultures in need of further workup.

In studies on blood cultures from children, Kahn et al [39] found 14.5% positive cultures out of 500 sampled by Bactometer 32 monitoring and 15% positive by conventional processing. The average detection time by Bactometer 32 was 8.5 h vs. 24 h by conventional analysis.

Kagan and associates working at the Clinical Centers of the National Institutes of Health [40] built their own impedance-measuring system to monitor blood cultures using an aliquot of a patient blood culture, which was first treated by a host cell lysis and filtration method. This links a very sensitive means of initiating culture of bacteria in blood with a sensitive system for detecting microbial metabolism and growth. The remaining patient blood culture was processed by conventional methods. In all, 53 blood cultures from 107 patients were positive by one or both methods. The lysis-filtration-impedimetry system detected 92% of the total number of positive cultures, compared with 56% detected by the conventional method. The impedance detector first detected all positives when there were about 5×10^5 bacteria per milliliter in the culture. This was detectable as a faint turbidity in the lysis-filtration culture, which was freed of erythrocytes. Lysis-filtration eliminated the need for a reference well and made the impedance change caused by microbial metabolism more readily detectable. The impedance effect of the patient cells masks this early detection.

Electrical impedance still offers great promise as a screening system for microbial growth in blood cultures, although it has not become established. It offers a combination of a sensitive detector and the possibility of continuous monitoring. A system that provides for the shaking of the culture, such as that described by the Wade-Honeywell patents, and an adequate number of channels will be necessary to produce a successful system. The system studied by Kagan and associates is an extremely sensitive detector. If the expense of labor and materials for the lysis-filtration step can be minimized, the early detection and high yield of positive cultures make this system very attractive.

Electrical conductivity in blood cultures was measured by Brown et al [41] using the Malthus 112L Microbiological Growth Analyzer. In a comparison of the Malthus system with conventional methods for examining blood cultures 100 clinically-significant, positive cultures were evaluated. Organisms grew from 82 of the conventional aerobic bottles, 78 of the conventional anaerobic bottles and 71 of the Malthus bottles. The Malthus system detected positive cultures sooner than the conventional methods in 83.6% of the positive cultures compared with 7.3% that were detected sooner by the conventional methods. A high false positive rate (26.9%) for the Malthus system was largely attributed to electrode instability. This problem may be overcome by recent modifications that have resulted in greater electrode stability [41,47].

B. Electrical Measurements to Screen for Bacteriuria

Urine cultures are the most frequent tests done by clinical microbiology laboratories. A significant proportion of the specimens will be free of bacteria. A screening method would be useful to select those specimens containing bacteria that require further work for identification and antimicrobial susceptibility testing and to estimate the number of bacteria they contain. Instruments that measure electrical impedance or conductance are capable of doing this screening and testing. As new screening methods are developed, it must be kept in mind that fewer than 10^5 bacterial per milliliter may be significant in urinary tract infection [42].

Wheeler and Goldschmidt [32] have described an unusual instrument for estimating the number of bacteria present in a specimen and have applied it to the analysis of urine (see Section II,E). In this analysis, urine is prefiltered through an 8-μm pore filter to remove host cells. Bacteria were washed on a small-pore membrane, resuspended in distilled water, and introduced into the impedance measuring cell. This is a direct measurement of bacterial cells without a culture stage. It is a direct outgrowth of the observation by Schwann [11,12] that biological materials in suspension exhibit capacitance and resistance changes as a function of applied frequency. Shape and size of bacteria did not influence the impedance. Correlation between the impedance and the final plate counts of viable cells was within 5%. The system was able to detect 10^3 and 10^4 bacteria per milliliter. This is a helpful capability in view of the significance of fewer than 10^5 bacteria per milliliter in urine [42]. This system has the potential for rapid analysis, but the many handling steps for preparing the specimen markedly increase technologist time over conventional methods.

Several groups have studied the Bactometer system for screening urine cultures. In a study of 200 clinical specimens, Throm et al [43] found an average detection time of 2.5 h for 41 clinically significant bacteria. The Bactometer detected α and nonhemolytic streptococci and "diphtheroids," which often do not appear on plate cultures until 48 h of incubation. Conventional cultures picked up 10^2 to 10^3 coagulase-negative staphylococci, yeasts, pseudomonads, or "diphtheroids," which were not detected at 20 h by the Bactometer. The continuous monitoring and recording system allowed the addition of cultures for analysis at any time and removal as detection of microbial activity was achieved. Zafari and Martin [44] studied 156 urine specimens of which 35 had colony counts greater than 10^5 organisms per milliliter. All urines with more than 10^5 organisms per milliliter were detected by decreased impedance within 4 h. Four false positives were found and 96.8% of 151 samples were correctly classified according to these criteria. A 2-h screen was nearly as efficient. Cady and et al [45] studied a screen for urine cultures with more than 10^5 organisms per milliliter. When an imped-ance-positive culture was defined as one that gave detectable impedance change within 2.6 h, 95.8% of 1,133 urine cultures were classified correctly as containing more than, or fewer than, 10^5 organisms per milliliter. Selection of a longer detection time decreases false negatives at the cost of increased false positives. False negatives by Bactometer studies may result from anti-biotic in the urine at an inhibitory level. Streaking onto plates may allow dilution of the antibiotic below the minimal inhibitory concentration, allow-ing colonial growth in parts of the plate with least antibiotic transfer. This is a common problem with urine specimens, because it is common medical practice to start some therapy for symptomatic urinary tract infection and only obtain a culture if symptoms persist. The Bactometer systems provide rapid detection times for a large proportion of specimens analyzed. The multiple channels available for analysis are essential for large laboratories. Only minimal setup time is required.

The Malthus conductimetric system was compared to the Bactometer system by Brown and Warner [46]. Very little difference was found in the results. The Bactometer uses a disposable module about ¼ the size of the Malthus nondisposable module. Antibiotics and preservatives may be diluted less in the Bactometer module. The use of 4- to 5-h cutoff times minimized false positives and false negatives for both systems. A 5-h screen unhappily is somewhat long for inclusion within the usual clinical work day.

C. Antimicrobial Susceptibility Testing

Measurement of antimicrobial susceptibility of bacteria requires an instru-ment that can detect the difference between the inoculum and a growing population of cells on one hand, and an inhibited population of cells on the other. Systems that measure impedance and possibly other electrical parameters are well adapted to this kind of testing. The multichannel analyses carried out by the Bactometer and the Orga 6 are especially well adapted to testing several concentrations of a number of different antimicro-bials. The usefulness of impedimetry for susceptibility testing has been suggested by several investigators [1,3,4,28,48,49,50]. In most cases, a 10^5 to 10^6 bacteria per milliliter inoculum is used so that the required distinction can be measured in some cases as early as 2-3 h. In a thorough study, Colvin and Sherris [48] compared readings by impedance change as compared to visual turbidity readings. They used a module of their own design, which contained 16 2-mL glass vials with stainless steel vertical electrodes. After overnight incubation and use of a 10^6 per milliliter inoculum, the impedance measurement was within one twofold dilution of the standard method 93% of the time. When the impedance end point was determined at 6h, the correlation was lowered to 34%. The greatest discrepancies occurred with ampicillin, tetracycline, and polymyxin E.

D. Reading Serological Tests

Tsuchiya et al [51] have studied the reading of serological tests by impedimetry. Impedance changes were marked during the red cell sedimentation of hemaglutination reactions and were also observed in hemolytic reactions and liposome sedimentation. Similar responses have been observed in studies of biological materials by Schwan [52], and were observed in early blood culture monitoring [3]. The changes were most marked when interdigitating flat electrodes were placed upon the bottom of the measuring module. Tsuchiya and co-workers [51] compared conventional reading of antistreptolysin O titers by the Rantz and Randall method and the results of an impedance measurement system. The correlation coefficient was 0.940. Further studies of impedimetric reading of serological end points are well justified. Improved standardization of end point readings and labor saving for laboratories with large serological test volumes could result.

E. Food Analysis

The detection and counting of microorganisms in foods has been used to evaluate food quality or keeping ability. The traditional techniques depend on bacterial growth to produce colonies or most probable number determinations. These methods require prolonged incubation before results can be read. As a consequence, microbiological testing has been retrospective. Tests are also very labor intensive. New and rapid analytical methods are being investigated. These methods would allow rapid grading of constituent materials and the final product. They could reduce storage space needs, allow products to move to the market more rapidly, and diminish labor costs. Impedimetry has become a highly successful system for rapid microbiological analysis of food. By use of special methods that allow growth of bacteria associated with spoilage or contamination to be noted, differential estimations of particular microorganisms is possible.

F. Milk

Martins et al [53] found that impedance measurements made with the Bactometer 32 on samples incubated at 21 and 32°C, provided a better means of estimating the shelf-life of pasteurized milk than the standard plate count and the psychrotrophic plate count. Eighty percent of the 61 samples were classified correctly by impedimetric measurement. The impedance method was also more rapid, requiring 14 h vs 2 d for the standard plate count and 10 d for the psychrotrophic plate count. In a later study by Bishop et al [54], the impedance method was compared with the Mosely test. The Mosely test has an adequate relationship to shelf-life, eg, a correlation coefficient of 0.84. The impedance detection time test incubated at 21 and 18°C had correlation coefficients of 0.88 and 0.87, respectively. The impedance detection time was also less labor intensive and required a day or two, vs 7-9 d for the Mosely test.

Gnan and Luedecke [55] studied the relationship of raw milk standard plate counts to impedance detection times measured on raw milk alone or raw milk treated with yeast extract. When a specific detection cut off time of 7 h was selected, and a maximum plate count of 10^5 CFU/mL was selected, 97% of the samples treated with yeast extract were classified correctly by impedance detection times. The readings at 7 h are preferable to the standard plate count reading requiring 48 h. Firstenberg-Eden and Tricarico [56] studied the relationship between impedance detection time and total, mesophilic (24 h at 37°C), and psychrotrophic plate counts. They found that 32°C incubation for the impedance assay was inappropriate for total count and that 18°C was a preferable temperature. At 32°C, most of the impedance change was due to mesophiles. These had an average generation time, or AGT, of 0.5 h, whereas psychrotrophs had an AGT of 3.0 h. Mesophiles, most often the predominant population when present at greater than 10^5 CFU/mL can be screened out at 4 h. Psychrotrophs were screened out at the same concentration within 21 h.

G. Analysis of Meat

Firstenberg-Eden [57] studied the use of impedance measurements for estimating the number of microorganisms in raw meat. A calibration curve could be constructed to relate approximate numbers of microorganisms per unit weight with the impedance detection time. Samples dispersed in brain-heart infusion medium with a stomacher served best for impedance assay. Frozen samples had a 1- to 2-h later impedance detection time for a similar plate count. This may result from stressing of bacteria by the freezing, resulting in a slower initial growth rate; likewise, frozen meat may contain different kinds of bacteria. An impedimetric most probable number estimate was developed by Martins and Selby [58] for the estimation of coliforms in meat.

H. Other Foods

Other food materials have been studied by impedimetry. Sorrells [59] found that estimates could be obtained for the bacterial content of grains by impedimetry. These were closely related to the usual plate count methods but faster. Hardy et al [60] found impedimetry to yield useful estimates of microbial concentration in frozen vegetables. Firstenberg-Eden and Klein [61] were able to improve the coliform detection by impedance assay using a new medium. This medium had better selectivity than lauryl tryptose broth and violet red bile agar (VRBA). Weihe et al [62] studied frozen concentrated orange juice by an impedimetric detection time assay. Using this assay more than 96% of 468 retail samples were classified correctly as having more than or less than 10^4 bacteria per milliliter.

I. Analysis of Effluent from Sewage Treatment Plants

Silverman and Munoz [63] devised an impedimetric technique for rapid estimation of coliforms in effluents from sewage treatment plants. Fecal coliforms growing in a selective lactose-based broth medium at 44.5°C generated a change in the electrical impedance of the culture relative to a sterile control when populations reached 10^6 to 10^7 per milliliter. A linear relationship was found between the log 10 of the number of fecal coliforms in an inoculum and the time required for an electrical impedance ratio signal to be detected. Detection times varied from 5.8-7.9 h for a sample containing 200 fecal coliforms to 8.7-11.4 h for a sample containing one fecal coliform. Variations in detection times were found for different sewage treatment plants. This difference may be related to the presence of inhibitors in the effluent. Oremland and Silverman [64] studied slurries of sediment from San Francisco Bay for their sulfate reducing capacity and ability to produce pure cultures of <u>Desulfovibrio aestuarii</u>. They found that the rate of sulfate reduction measured by $^{35}SO_4$ correlated with the rate of change in electrical impedance ratio with time.

IV. CONCLUSIONS

A highly sophisticated technology has evolved for impedimetric analysis of food products, and more recently for quality control of health and beauty products. Calibration curves can be constructed to relate impedance detection times to standard plate counts at different temperatures. In most cases, less effort in making dilutions and counting plates is required in the impedimetric methods. Impedimetric methods are usually more rapid, so that materials need not be stored waiting for quality control results and final product can be shipped more rapidly. These new assays have been accepted by the American Public Health Association. Manufacturers show interest because of the savings in time and money for them. Improvements in instrumentation have included computer-assisted analysis of data, improved hard copy, and video display of data. These same types of assay are being studied for sewage treatment effluent.

Although much of the initial study of electrical impedance assay systems involved clinical assays, this has not continued to be so. This speaks more about the cost of development for clinical use than about the adequacy of electrical impedance as an assay method. The continuous monitoring capacity of electrical impedance allows screening to be done on a timely basis whenever a specimen is received for analysis. Improved electrode assemblies and the capacity to speed growth of microorganisms by shaking have been patented but not used in commercial instruments.

Rapidity of measurement is stressed in new instrumental systems. To a degree rapidity of measurement is dependent upon the sensitivity of the instrument for detecting small numbers of microorganisms. However, just as important may be the improvement of culture systems so that microorganisms can grow more rapidly to their detection threshold.

V. REFERENCES

1. Ur, A.; Brown, D.F.J. In "New Approaches to the Identification of Microorganisms"; Heden, C.; Illeni, T. eds.; John Wiley and Sons: New York, 1975; pp. 61-71.
2. Cady, P. In "New Approaches to the Identification of Microorganisms"; Heden, C.; Illeni, T., eds.; John Wiley and Sons: New York, 1975; pp. 73-99.
3. Hadley, W.K.; Senyk, G. In "Microbiology 1975"; Schlessinger, D., ed.; American Society for Microbiology: Washington, DC, 1975; pp. 12-21.
4. Cady, P. In "Mechanizing Microbiology"; Sharpe, A.N.; Clark, D.S., eds.; Thomas: Springfield, IL, 1978; pp. 199-239.
5. Cady, P.; Dufour, S.W.; Shaw, J.; Kraeger, S.J. J. Clin. Microbiol., 1978, 7, 265.
6. Eden, R.; Eden, G. "Impedance Microbiology"; Research Studies Press: Letchworth, Herts, SG 63BE England, and John Wiley and Sons, New York, 1984.
7. Gall, L.S.; Curby, W.A. "Instrumental Systems for Microbiological Analysis of Body Fluids"; CRC Press Inc., Boca Raton, FL, 1980.
8. Lawrence, A.J.; Morres, G.R. Eur. J. Biochem., 1972, 24, 538.
9. Hans, M.; Rey, A. Biochim Biophys Acta, 1971, 227, 618.
10. Richards, J.C.S.; Jason, A.C.; Hobbs, G.; Gibson, D.M.; Christie, R.H. J. Phys. E, Sci. Instr. (London) 1978, 11, 560.
11. Schwan, H.P. Biophysik, 1966, 3, 181.
12. Schwan, H.P. In "Physical Techniques in Biological Research", Vol. 6, Nastuk, W.L., ed.; Academic Press, New York, 1963, pp. 323-407.
13. Firstenberg-Eden, R.; Zindulis, J. J. Microbiol. Meth., 1984, 2, 103.
14. Eden, G.; Eden, R. IEE Trans. Biomed. Eng., 1984, BME-31, 193.
15. Geddes, L.A.; Baker, L.E. In "Principles of Applied Biomedical Instrumentation"; John Wiley and Sons, New York, 1968, pp. 150-205.
16. Stacy, R.W. "Biological and Medical Electronics"; McGraw, New York, 1960, Chapter 3.
17. Stewart, G.N. J. Exper. Med., 1899, 4, 235.
18. Oker-Blom, M. Z. Bakteriol., Abt. 1, 1912, 65, 382.
19. Parsons, L.B.; Sturges, W.S. J. Bacteriol., 1926, 11, 177.

20. Parsons, L.B.; Sturges, W.S. J. Bacteriol., 1926, 12, 267.
21. Parsons, L.B.; Drake, E.T.; Sturges, W.S. J. Bacteriol., 1929, 51, 166.
22. Sierakowski, S.; Leczycka, E. Z. Bakteriol. Parasitenkd. Infektionskr. Hyg., Abt. 1, Reihe B, 1933, 127, 486.
23. Allison, J.B.; Anderson, J.A.; Cole, W.H. J. Bacteriol., 1938, 36, 571.
24. McPhillips, J.; Snow, N. Aust. J. Dairy Technol., 1958, 3, 192.
25. Ur, A. Nature (London), 1970, 226, 269.
26. Ur, A. Biomed. Eng., 1970, 5, 342.
27. Ur, A. Am. J. Clin. Pathol., 1971, 56, 713.
28. Ur, A.; Brown, D.F.J. J. Med. Microbiol., 1974, 8, 19.
29. Ur, A.; Brown, D.F.J. Bio-med. Eng., 1974, 9, 18.
30. Wade, G. J. (Honeywell Inc.). U.S. Patent 4,250,266 (1981).
31. Wade, G. J. (Honeywell Inc.). U.S. Patent 4,267,276 (1981).
32. Wheeler, T.G.; Goldschmidt, M.C. J. Clin. Microbiol., 1975, 1, 25.
33. Wilkins, J.R.; Stoner, G.E.; Boykin, E.H. Appl. Microbiol., 1974, 17, 949.
34. Lamb, V.A.; Dalton, H.P.; Wilkins, J.R. Am. J. Clin. Pathol., 1976, 66, 91.
35. Walkins, J.R. Appl. Environ. Microbiol., 1978, 36, 683.
36. Specter, S.; Throm, R.; Strauss, R.; Friedman, H. J. Clin. Microbiol., 1977, 6, 489.
37. Buckland, A.; Kessock-Philip, S.; Bascomb, S. J. Clin. Pathol., 1983, 36, 823.
38. Hadley, W.K. In "Second International Symposium on Rapid Methods and Automation in Microbiology"; Cambridge, U.K.; Learned Information Ltd.: Oxford, 1976; p. 14.
39. Kahn, W.; Friedman, G.; Rodriguez, W.; Controni, G.; Ross, S. In "Second International Symposium on Rapid Methods and Automation in Microbiology", Cambridge, U.K.; Learned Information Ltd.: Oxford, 1976; pp. 14-15.
40. Kagan, R.L.; Schuette, W.H.; Zierdt, C.H.; MacLowry, J.D. J. Clin. Microbiol., 1977, 5, 51.
41. Brown, D.F.J.; Warner, M.; Taylor, C.E.D.; Warren, R. E. J. Clin. Pathol., 1984, 37, 65.
42. Stamm, W.E.; Counts, G.W.; Running, K.R.; Fihn, S.; Turck, M.; Holmes, K.K. New Engl. J. Med., 1982, 307, 463.
43. Throm, R.; Specter, S.; Strauss, R.; Friedman, H. J. Clin. Microbiol., 1977, 6, 271.
44. Zafari, Y.; Martin, W.J. J. Clin. Microbiol., 1977, 5, 545.
45. Cady, P.; Doufour, S.W.; Lawless, P.; Nunke, B.; Kraeger, S.J. J. Clin. Microbiol., 1978, 7, 273.
46. Brown, D.F.J.; Warner, M. In "Proc. 3rd International Symposium on Rapid Methods and Automation in Microbiology"; Washington, DC, pp. 171-175.
47. Baynes, N.C.; Comrie, J.; and Prain, J.H. Med. Lab. Sci., 1983, 40, 149.
48. Colvin, H.J.; Sherris, J.C. Antimicrob. Agents Chemother., 1977, 12, 61.
49. Tsuchiya, T.; Terashima, E. Nihon Univ. J. Med., 1978, 20, 319.

50. Tsuchiya, T.; Terashima, E.; Kawano, K.; Nakano, E.; Takenaka, M.; Kumasaka, K.; Tsuchiya, T. Nichidai Igaku Zasshi, 1978, 37, 405.
51. Tsuchiya, T.; Hasimoto, M.; Kawano, K.; Terashima, E.; Kumasaka, K.; Tsuchiya, T. Nichidai Igaku Zasshi, 1980, 39, 31.
52. Schwan, H.P. Ann. N.Y. Acad. Sci., 1968, 148, 191.
53. Martins, S.B.; Hodapp, S.; Dufour, S.W.; Kraeger, S.T. J. Food Protect., 1982, 45, 1221.
54. Bishop, J.R.; White, C.H.; Firstenberg-Eden, R. J. Food Protect., 1984, 47, 471.
55. Gnan, S.; Luedecke, L.O. J. Food Protect., 1982, 45, 4.
56. Firstenberg-Eden, R.; Tricaro, M.K. J. Food Sci., 1983, 48, 1750.
57. Firstenberg-Eden, R. Food Technol., Jan. 1983, 64.
58. Martins, S.B.; Selby, M.J. Appl. Environ. Microbiol., 1980, 39, 518.
59. Sorrells, K.M. J. Food Protect., 1981, 44, 832.
60. Hardy, D.; Kraeger, S.J.; Dufour, S.W.; Cady, P. Appl. Environ. Microbiol., 1977, 34, 14.
61. Firstenberg-Eden, R.; Klein, C.S. J. Food Sci., 1983, 48, 1307.
62. Weihe, J.L.; Seibt, S.L.; Hatcher, W.S., Jr. J. Food Sci., 1984, 49, 243.
63. Silverman, M.P.; Munoz, E.F. Appl. Environ. Microbiol., 1979, 37, 521.
64. Oremland, R.S.; Silverman, M.P. Geomicrobiol. J., 1979, 1, 355.

INDEX